T0201868

Life Without Degrees of Moral Status

Life Without Degrees of Moral Status

Implications for Rabbits, Robots, and the Rest of Us

DAVID S. WENDLER

OXFORD
UNIVERSITY PRESS

Oxford University Press is a department of the University of Oxford. It furthers the University's objective of excellence in research, scholarship, and education by publishing worldwide. Oxford is a registered trade mark of Oxford University Press in the UK and certain other countries.

Published in the United States of America by Oxford University Press
198 Madison Avenue, New York, NY 10016, United States of America.

Library of Congress Cataloging-in-Publication Data
Names: Wendler, David, author.
Title: Life without degrees of moral status : implications for rabbits,
robots, and the rest of us / David S. Wendler.
Description: New York, NY : Oxford University Press, [2023] |
Includes index. Identifiers: LCCN 2022060630 (print) | LCCN 2022060631 (ebook) |
ISBN 9780197675328 (hardback) | ISBN 9780197675342 (epub)
Subjects: LCSH: Human-animal relationships—Moral and ethical aspects. |
Animal welfare—Moral and ethical aspects.
Classification: LCC QL85 .W37 2023 (print) | LCC QL85 (ebook) |
DDC 590—dc23/eng/20230106
LC record available at https://lccn.loc.gov/2022060630
LC ebook record available at https://lccn.loc.gov/2022060631

DOI: 10.1093/oso/9780197675328.001.0001

Printed by Integrated Books International, United States of America

Disclaimer: The present work was supported by intramural research funds at the US NIH Clinical Center. However, the views expressed are the author's own. They do not represent the position or policy of the Clinical Center, NIH, DHHS, or US government.

Disclaimer: The present work was prepared by [illegible] research [illegible] US NIH [illegible] the views [illegible] NIH [illegible]

OOOOOOOOOOMMMMMMMMMM

Contents

Acknowledgments

The writer Raymond Chandler once described himself as a fellow who "writes 30,000 words to turn in five." My experience with this book hasn't been that extreme, but it hasn't been too far off either. Credit for the time readers will save goes to the friends, colleagues, and reviewers who urged me to trim it down. They read earlier, longer, and less clear versions, while I get credit for the final, shorter, and, hopefully, clearer one. To them, my deepest thanks, for comments and recommendations, minor and major, recent and long ago: participants at the NIH Bioethics Joint Bioethics Colloquium, the New York University Bioethics Colloquium, the NIH reading group, Sean Aas, Bernardo Aguilera, David Benatar, David DeGrazia, Jake Earl, Hope Ferdowsian, Sam Garner, John Gluck, Tyler John, Rebecca Johnson, Holly Kantin, Matthew Liao, Frank Miller, Joe Millum, Collin O'Neil, Tina Rulli, Ben Sachs, Ben Schwan, Jeff Sebo, Robert Steel, Camilla Strassle, Connor Sullivan, David Wasserman, Alan Wertheimer, and an anonymous reviewer for Oxford University Press.

Introduction

Why Should We Care?

I.1. Reasons to Care

Going to the beach, and playing in the sand, offers a rare opportunity for children to let their energies, creative and destructive, run wild without parents having to worry that they might do something wrong. As prizes in sand castle competitions attest, there can be aesthetically better and worse ways to arrange grains of sand. But there aren't morally better and worse ways. Parents don't have to worry that their children might treat the sand inappropriately.

Of course, children can do inappropriate things *with* grains of sand. Building a sand castle and smashing it to smithereens is not a matter for moral concern. Throwing the remains in the face of a baby is. This difference between what children do *to* grains of sand versus what they do *with* grains of sand points to one of the fundamental distinctions in our lives: some things matter morally; other things don't. Children matter morally; they have moral status. Grains of sand don't matter morally; they don't have moral status.

We will briefly consider why this is the case—why some things have moral status and other things don't. But, for the most part, we will simply assume that it is in order to explore a related and critically important question: Among all the things that have moral status, do some things matter more than others? In other words, does moral status come in degrees? Many people think so. They think, for example, that animals matter morally, but human beings matter much more (see Figure I.1: Degrees of Moral Status I).[1] Some people think

Life Without Degrees of Moral Status. David S. Wendler, Oxford University Press.
DOI: 10.1093/oso/9780197675328.003.0001

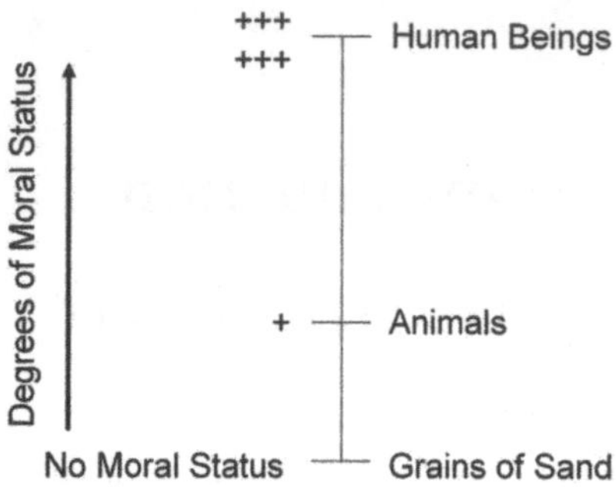

Figure I.1 Degrees of Moral Status I

this in an explicit and conscious way. They have taken the time to think about the moral status of animals and human beings, and they have concluded that ours is higher than theirs. Others have never really thought about it. Still, they mildly scold their child for hitting the neighbor's cat, whereas they punish their child for hitting the neighbor. They do this because they think, often without realizing it, that human beings are morally much more important than animals.

This belief shapes our lives in ways both trivial and profound. Consider throwing a party for more people than there are seats in the house. Some will get a seat; others will have to stand or sit on the floor. Where does the cat go? The cat takes up less space, but it leaves more hair. The cat is less likely to have an interesting story to share, but it is in greater danger of being stepped on. Taking into account all the differences between human beings and cats, and where they sit, would take more time than most parties last. Yet making these decisions takes hardly any time at all: we simply chase the cat off the sofa and give the spot to the human.[2] Why? Because we think human beings are morally much more important than cats. For the same reason, we treat animals in ways that we would never treat human beings: we put animals in cages, use them in pain-inducing experiments to develop new shades of lipstick, and eat them for dinner.

Our discussion will focus on *sentient* animals, that is, animals that have experiences which can be good or bad for them. Animals that experience pain and pleasure, anxiety and excitement, including cats, sloths, rabbits, great blue herons, alligators, pigs, Amazon river dolphins, and chimpanzees. Sentient animals, as this list illustrates, are very different from us, and from one another. These differences explain why the final conclusion of the present work—there are no degrees of moral status—does not imply that animals should be treated like human beings, or that all sentient animals should be treated alike. Taking a ride on the back of a horse may be acceptable; doing the same thing to a poodle is not. Granting the importance of these differences for determining how animals ought to be treated, much of the present discussion will focus on rabbits. This is not because rabbits are especially important, nor that how we should treat rabbits reflects how we should treat all sentient animals. Instead, the goal is to simplify the discussion by focusing on a sentient animal people are familiar with without biasing the conclusions by appealing to animals many people are attached to (e.g., dogs or cats).

The importance of whether there are degrees of moral status is not limited to sentient animals, however. It is critical to how we treat infants and people with Alzheimer disease. It is also critical to how we should treat robots and genetically enhanced human beings, and how they should treat us. The belief that there are degrees of moral status justifies our treating human beings better than animals. But it also raises the possibility that robots and genetically enhanced human beings could become morally more important than we are. If that happens, it might be ethical for them to put us in cages, to subject us to painful experiments for their benefit, and to eat us for dinner. These possibilities provide important reasons for us to care, for us to take the time to figure out whether, in fact, there are degrees of moral status among those who matter morally. Are we morally more important than animals? Could robots and genetically enhanced human beings become morally more important than the rest of us?

If it turned out that human beings were morally more important than animals, the present discussion might be of theoretical interest, but it would have little practical importance, beyond perhaps confirming what many of us already believe and how we already behave. In fact, the present discussion concludes that there are no degrees of moral status among those who matter morally. This conclusion, as we shall see, is consistent with many of our existing beliefs and practices. For example, even though there are no degrees of moral status, it is still significantly worse to kill human beings than animals, and it is more important for humans to direct the course of their own lives. But this conclusion also reveals that a number of our current beliefs and practices are mistaken. It is not worse, as many people think, to cause human beings to experience pain and to suffer than it is to cause animals to experience pain and to suffer to the same extent. And respect for individuals does not apply only to human beings; it applies to animals as well. These implications reveal that we will need to rethink and revise, in important ways, how we treat animals.

I.2. Examples and Intuitions

To find out what the world is like, we go outside and look. We look around us, we look into the sky, and we look into microscopes. We can similarly learn a good deal by looking at human beings and rabbits. Which one is taller? Which one has a tail? But this won't tell us whether human beings have higher moral status than rabbits. Moral status is not something we can discover by looking. So how do we figure out whether human beings are morally more important than animals? We start, as discussions in ethics typically do, with our reactions to specific cases, with our intuitions. I describe a case and ask you to consider what you ought to do. Your response, along with the responses of others, provides evidence for what morally is right and what morally is wrong. Let's begin, then, with two examples:

> **Experimentation**: You find yourself on a committee that reviews medical research. A researcher comes to the committee with a proposal to test a potential new treatment for lymphoma, a type of cancer. The study will cause significant pain to five subjects for four hours, and there is no chance that being in the study will benefit the subjects. The amount of pain and the scientific value of the study will be the same whether the researcher enrolls human beings or rabbits. Who should the researcher enroll?

Before you answer, notice that there are ways in which it would be better to enroll human beings. They can consent to being enrolled; rabbits can't. Humans who enroll might learn something about medical research, and they might experience some pride in contributing to it. Rabbits can't do that either. Nonetheless, most people think it is ethically better to conduct risky and pain-inducing experiments with animals rather than humans. Indeed, as we will consider in section 8.5, regulations around the world *require* research on potential new medical treatments to start with animals. This is because existing regulations assume that we are morally much more important; hence, we deserve greater protection than animals: "Most of us regard it as clear that ordinary adult human beings have higher moral status than non-human animals."[3] Willard Gaylin, past president of the Hastings Center, one of the most important centers for bioethics in the world, said something similar: "Respect for human beings demands a dignity granted our species that is beyond any qualitative comparison with others. We are sui generis."[4] The moral philosopher Baruch Brody identified and then endorsed the implications of this view for pain-inducing experiments: "the same unit of pain counts less morally if experienced by an animal than it would if experienced by a human being."[5] This brings us to the second example:

> **House on Fire**: While taking a walk one day, you come upon a (literal) house on fire. A child runs out and tells you that there are two individuals trapped inside. In the room to the left there is a rabbit and in the room to the right there is a human being. You have time to run in and go left or right, but not both. What should you do?

Almost everyone I know has several reactions. First, the fact that the rabbit is likely to experience pain and to suffer gives you some reason to rescue it. This supports the claim that sentient animals have moral status. Second, you should save the human being, even though the human being and the rabbit will suffer to the same extent.

Third, this seems obvious, even though you don't know much about the human being, or the rabbit. You don't know whether the human being is fifteen or ninety-five. You do not know whether they are nice or mean. All you know is that one individual is a human being and the other individual is a rabbit, and that is enough to know what you ought to do: you should save the human being because human beings are more important morally.

Fourth, human beings are not just somewhat more important; we are significantly more important. To see this, imagine you come on the scene a few minutes after someone else faced the same dilemma. They turn to you, explain the choice they faced, and show you the rabbit they just rescued. Would you think: *Oh well, it was a close call.* No. You would think they had made a terrible mistake: they should have saved the human being. This suggests human beings are not just somewhat more important; we are significantly more important. In this way our intuitions support the claim that there is a "significant moral divide" between people and animals.[6]

Our reactions to *Experimentation* and *House on Fire* provide some reason to think there are degrees of moral status among those who matter morally. They also illustrate two ways to understand this difference. The first way, illustrated by Baruch Brody's quote, claims that the same amount of pain or suffering is morally worse when it occurs in human beings. This understanding focuses on the *experiences* of individuals who have moral status. If human beings have higher moral status, then our experiences matter more than the experiences of rabbits. It is more important for us to have positive experiences, and it is worse for us to have negative ones. The second way to understand differences in degrees of moral status focuses on how we *treat* others; it focuses on the actors rather than the subjects of our actions. If human beings have higher moral status, we have stronger reasons to treat human beings well than to treat rabbits well.

Because these are both common ways to understand differences in degrees of moral status, we will consider both of them. Sometimes I will talk about differences in moral status in terms of the experiences of individuals. On this understanding, *greater* moral status involves human beings' interests in having good experiences and avoiding bad experiences mattering more than the analogous interests of animals. It involves our pain, anxiety, and suffering, our pleasure and happiness, mattering more. I will also talk about greater moral status in terms of individuals having claims or rights to be treated with respect, for example, claims or rights to direct the course of their lives and to not be sacrificed for the benefit of others.

The common belief that we are morally more important than rabbits raises the question of how many levels of moral status there might be (see Figure I.2: Degrees of Moral Status II).[7] Consider a variation on *Experimentation*. This time you have a choice between doing the experiment with rabbits or chimpanzees. Which would you choose? Consistent with a 2015 decision by the US National Institutes of Health to stop funding research on chimpanzees, while

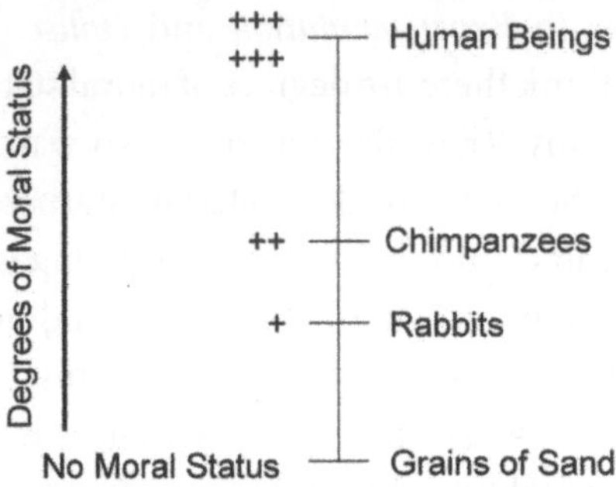

Figure I.2 Degrees of Moral Status II

continuing to fund research on other animals,[8] many people think it is worse to cause pain to chimpanzees than rabbits.[9] To assess this view, we will need to determine whether there are degrees of moral status and, if so, how many there are. Do all sentient animals have the same level of moral status? Or are some sentient animals more important than others; for example, are chimpanzees more important morally than rabbits?

To answer these questions, we will consider the possible basis for our (presumed) greater moral status: What might account for the fact that we have higher moral status than rabbits? If we can figure that out, we can use the answer to assess whether different types of animals have the same or different levels of moral status.

I.3. How Many Levels Are There?

Proponents assume that, if we have higher moral status than rabbits, it is because we possess some property or properties which they lack, or which they don't possess to the same extent (in what follows, I will not always repeat this second possibility). To see why, imagine someone claiming that human beings just happen to be morally more important than rabbits. There are, they claim, no differences between human beings and rabbits that explain this difference; it is just a fact. Or imagine they claim to be morally more important than *you* are, even though there are no morally relevant differences between the two of you which might explain this difference. That would be odd. Some individuals don't just happen to be morally more important than others. Higher moral status isn't something that just happens to attach to some individuals, but not others.

If some individuals are morally more important than others, there must be something that explains this difference; there must be some property or properties which they possess that increase their level of moral status.[10] Moreover, the property or properties, whatever they might be, are likely to be important or significant in their own right. To see why, consider what Charles Darwin took to be the most characteristic of all human expressions: blushing. Imagine someone arguing that we are morally more important than rabbits because we blush and they don't. That, too, would be odd. The fact that we blush can't make it morally worse to cause us to experience pain and to suffer in medical experiments compared to rabbits, and it can't give us rights that rabbits lack. If we are morally more important than rabbits, it must be because we possess some important property or properties which they do not possess (or they possess to a lesser extent).

The properties which could make us more important morally come in degrees. Here, for example, is a common view of moral status: sentience gives individuals moral status so that individuals who are sentient have moral status (e.g., rabbits), while individuals who lack sentience do not have moral status (e.g., grains of sand). To this, proponents of degrees of moral status add that possessing one or more "superior" cognitive capacities increases individuals' moral status, with the result that human beings have higher moral status than sentient animals that do not possess the cited cognitive capacities (in section 8.6 we will consider the implications of this view for human beings who lack the cited cognitive capacities). Some proponents argue that being moral agents, that is, having the ability to recognize and act on the principles of morality, endows human beings with higher moral status than rabbits (for a listing of commonly cited superior cognitive capacities, see Appendix). On this view, rabbits have baseline moral status because they are sentient, while we have higher or "full" moral status because we are sentient and we are also moral agents. The interests of rabbits thus matter morally while our interests matter more than the analogous interests of rabbits and/or we have rights to be treated with respect that rabbits lack. As we go along, it will be important to remember that, although moral status is valuable, it is not the only way in which individuals can be valuable. Victoria Falls, Machu Picchu, the Mona Lisa, and the Magna Carta all are valuable, but none of them has moral status (section 6.3 considers why not). Instead, they matter in other ways: they have natural, aesthetic, cultural, and/or historical value.

There are two ways in which our superior cognitive capacities could increase our moral status, moving us from baseline moral status to higher or full moral status. Some proponents think that the fact that average adults possess superior cognitive capacities implies that *all* human beings have higher or full moral status. As we shall see, this understanding faces the challenge of explaining why human beings who do not themselves possess superior cognitive capacities nonetheless have full moral status. Why does the fact that average adults possess superior cognitive capacities imply that infants, who don't, still possess full moral status?

Other proponents avoid this challenge by maintaining that higher or full moral status applies only to the specific individuals who possess superior cognitive capacities (sometimes called persons). This includes competent adults, as well as any other individuals, perhaps some aliens, who possess superior cognitive capacities. This understanding of degrees of moral status raises concern that, if possession of superior cognitive capacities is the only way to increase individuals' moral status beyond the baseline, it could turn out that humans who lack superior cognitive capacities have the same moral status as rabbits. This is a significant concern: infants do not possess superior cognitive capacities and adults with Alzheimer disease, estimated to be fifty million people worldwide and growing, lose their superior cognitive capacities over time. The claim that possession of superior cognitive capacities significantly increases individuals' moral status thus justifies how we treat rabbits. But it raises the possibility that, at the beginning of our lives, all of us had the moral status of rabbits and, at the end of our lives, hundreds of millions of us will again. We will discuss this challenge in section 8.6.

Both of these ways of understanding degrees of moral status (all human beings have higher moral status; only individuals who possess superior cognitive capacities have higher moral status) assume that possession of one or more superior cognitive capacities increases individuals' moral status. The present discussion of whether some properties can do this thus applies to both views. And the final conclusion that none can reveals that both views are mistaken. To simplify the discussion though, I will frequently focus on the claim that all human beings have higher or full moral status, leaving implicit the fact that the discussion also applies to the claim that only individuals who possess superior cognitive capacities (persons) have higher or full moral status.

To briefly consider another example, which we will discuss in section 8.7, injecting human cells into the embryos of animals makes it possible to develop animals with human characteristics, so called animal-human chimeras. If our greater moral status traces to one or more superior cognitive capacities, using these techniques to give animals similar cognitive capacities might increase their level of moral status. Giving them enough might result in their moral status being equal to ours, raising the question over whether we should permit researchers to even initiate this line of research.

The question of whether there are degrees of moral status is also important for any individuals who have significantly greater cognitive capacities than average human adults. Computers can already

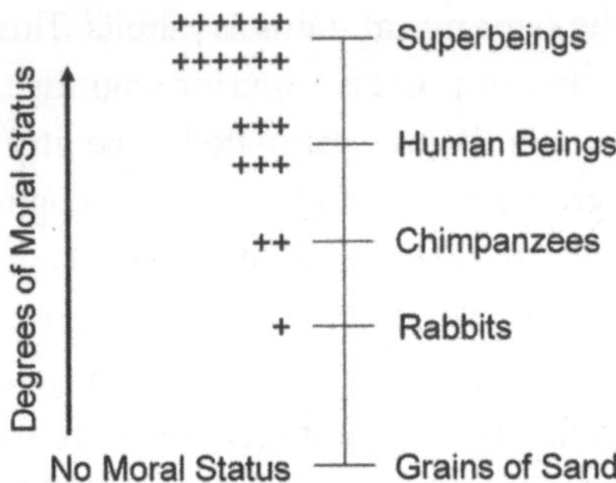

Figure I.3 Degrees of Moral Status III

do many things better than we can. Imagine the day when they are able to do essentially everything better, significantly better. Or consider a time when technology has advanced to the point where trillionaires can pay for gene editing, which makes their children cognitively superior to us and ours. If there are degrees of moral status, and individuals' level of moral status is determined by the extent to which they possess superior cognitive capacities, these individuals might turn out to have significantly greater moral status than we do (see Figure I.3: Degrees of Moral Status III). If that happens, they might be morally permitted to put us in cages, to use us in their pain-inducing experiments, and to eat us for dinner. Before we begin research projects that could lead to genius robots or superbabies, it would therefore be nice to know: Are there degrees of moral status among those who matter morally?

I.4. The Approach We Will Take

We began with our intuitions in response to two examples, *Experimentation* and *House on Fire*. The intuitions that it is better to do the experiment with rabbits, and to save the human being, suggested that we are morally more important than rabbits. And this suggested that, among all the individuals who matter morally, some matter more than others; in other words, there are degrees of moral status. So how do we assess whether these intuitions, and the views they point to, are right?

While our intuitions offer a good place to start, simply endorsing whatever they tell us is a bad place to end up. Some of our intuitions do not reflect what is true; they reflect our biases or, as we shall see in Chapter 1, our perspective on the world. Moreover, not everyone has the same intuitions. Some people do not think it is morally better to conduct the pain-inducing experiment in rabbits than human beings; hence, reliance on these intuitions alone leads to a stalemate. To try to resolve it, we will consider how degrees of moral status, if they existed, would work. What would have to be true for us to have greater moral status than rabbits, and are those things true? A finding that they are would support the claim that there are degrees of moral status. A finding that they aren't would undermine it.

When I first started this project, I assumed, based on my own responses to examples like *Experimentation* and *House on Fire*, that there are degrees of moral status. But the more I thought about it, the less confident I became, until finally I concluded that there are no degrees of moral status. This book tries to explain why. Very briefly, degrees of moral status require what I will call "moral status enhancing" properties, properties which make the individuals who possess them more important morally than other individuals who have moral status but don't possess the cited properties (at least to the same extent). I mentioned earlier that many proponents of degrees of moral status maintain that our superior cognitive capacities make us morally more important than rabbits. I am not going to suggest, in response, that we do not possess superior cognitive capacities, or that rabbits do. Instead, I will argue that this difference does not make us more important morally for the simple reason that there are no properties which increase individuals' moral status compared to individuals who have moral status but don't possess the cited properties. From this, the primary conclusion of the book follows: there are no degrees of moral status. Either you have moral status or you don't. And all those who possess moral status possess it to the same degree.

Given that degrees of moral status are endorsed by most major moral theories, I will not take a stand on which theory is right. Instead, I will try to assess whether, on any plausible moral theory, there are degrees of moral status. While this approach will make the present assessment more comprehensive, it also will make it somewhat more complicated. Proponents of different moral theories disagree over exactly what having greater moral status amounts to. Some understand it in terms of a difference in rights: our presumed greater moral status involves our possessing rights, such as a right to life and a right to direct the course of our lives, which rabbits lack. Others understand differences in moral status in terms of the moral importance of the individuals' experiences or interests. On this view, our greater moral status implies that it is worse for us to experience pain than it is for rabbits to experience the same level of pain. To assess whether there are degrees of moral status without limiting the discussion to a particular moral theory, I will understand degrees of moral status in these different ways, in terms of a difference in individuals' morally relevant claims, rights, interests, duties, and/or obligations. To simplify the discussion though, I will frequently refer to just one of these views at a time, with the assumption that the discussion applies to the other versions as well.

The final conclusion that there are no degrees of moral status appears to conflict with the intuitions many of us have. It conflicts with the intuition that it is better morally to conduct pain-inducing experiments in rabbits than humans. It conflicts with the intuition that, if you can save only one individual from the *House on Fire*, you should save the human being. The present conclusion thus leaves us with a challenge: Either it is mistaken or our intuitions in these cases are leading many of us astray. To assess which it is, we will reconsider the cases in detail, in Chapters 4, 5, and 6, and our responses to them.

I.5. A Brief Overview

Chapter 1 provides important background by explaining the nature of moral status and degrees of moral status. It also explains the difference between having more at stake versus having higher moral status. Chapter 2 considers the claim that we can dismiss the possibility of degrees of moral status out of hand. After rejecting this claim, Chapter 2 explains the difference between moral status conferring properties, moral action guiding properties, and moral status enhancing properties, arguing that degrees of moral status require moral status enhancing properties. Chapter 3 then considers whether any properties are moral status enhancing. It concludes that there are no moral status enhancing properties and, hence, no degrees of moral status. You either have moral status or you don't. In light of this conclusion, Chapter 4 considers whether it really is worse, as most of us believe, to kill human beings than animals, Chapter 5 considers whether aversive experiences, such as pain and suffering, really are worse when they occur in human beings than in animals, and Chapter 6 considers whether respect really applies to human beings but not animals.

Chapter 7 considers two arguments which claim that, even though there are no degrees of moral status, we should believe in them anyway. The first claims that believing in degrees of moral status offers a valuable rule of thumb which increases the chances that we will act appropriately, while the second claims that proper partiality gives human beings reason to favor other human beings over animals. The conclusion that these views are mistaken—hence, we should accept the fact that there are no degrees of moral status—raises many questions and challenges. Chapter 8 considers seven of the more important ones: (1) What's the difference between a world that includes degrees of moral status and one that doesn't? (2) Do we have to spend all our time helping animals? (3) Do animals have rights? (4) What are the ethics of raising animals and eating them? (5) What are the ethics of animal experimentation? (6) What are the implications for humans who lack superior cognitive capacities, and (7) What are the implications for chimeras, robots, and enhanced human beings?

All of these questions have been addressed by writers before us. The goal of the present work, however, is not to summarize their views and decide which one or ones are right. Instead, the goal is to consider what implications the conclusion that there are no degrees of moral status has for these questions. Which views does this conclusion preclude? Which ones does it support? What questions will we need to answer to develop a full understanding of a world without degrees of moral status? For these purposes, I will pick and choose from among the discussions that have preceded us, falling far short of giving those who have worked on these issues, or the views they endorse, their due. For readers who are interested in learning more about them, I provide references along the way.[11]

1
Our Place in the Universe and in the Analysis

Introduction

John Stuart Mill, a famous English philosopher, once claimed that it is better to be an unhappy human being than a happy pig (for a related discussion, see Box 1.1: What Is It Like to Be a Bat?). While that seems right, how could we make sure that it is? Presumably, we would do something like the following: We would imagine being a human being who is unhappy, imagine being a pig that is happy, and compare the two. Or we might imagine a genie who offers to turn us into very happy pigs for the rest of our lives. The fact that most of us would turn her down suggests Mill was right: it is better to be a human being, with all life's ups and downs, than a happy pig. But there's a catch.

To make this comparison, we were supposed to imagine what it is like to be a happy pig, and that is what we tried to do. But it isn't, in fact, what we did. Instead, we imagined what it would be like *for us* to be happy pigs. We imagined ourselves in a pig's body, living a happy pig's life. That seems terrible. But that isn't what it is like to be a happy pig. Happy pigs don't have our minds, our thoughts, our preferences. They have theirs. To assess what their lives are like, we would need to imagine what it is like for a pig to be a pig; we would have to imagine the life of a pig from the inside, from the perspective of a pig. Or we would need to evaluate the lives of human beings and pigs from some neutral perspective, neither ours nor

Life Without Degrees of Moral Status. David S. Wendler, Oxford University Press.
DOI: 10.1093/oso/9780197675328.003.0002

Box 1.1 What Is It Like to Be a Bat?

The fact that animals have moral status implies that we should consider the impact on them when deciding what to do. For example, the moral appropriateness of placing drops of chemicals in the eyes of rabbits to test different medications and cosmetics depends, in part, on what it feels like for the rabbits. And, to figure that out, we need to figure out what it is like for rabbits to have that experience. The philosopher Thomas Nagel posed a similar challenge: What is it like to be a bat: "If I try to imagine this, I am restricted to the resources of my own mind, and those resources are inadequate to the task. I cannot perform it either by imagining additions to my present experience, or by imagining segments gradually subtracted from it, or by imagining some combination of additions, subtractions, and modifications."[19]

Bats are interesting in this regard because they navigate, not by sight as we do, but by sound waves. This raises the possibility that their experiences of the world, how the world seems to them, might be very different from how it seems to us. If we can't figure out what it is like to be a bat, or even a rabbit, how can we figure out how we should treat them? How can we determine which actions cause animals to suffer, and to what extent, and which ones don't? How do we determine what proper respect involves when it comes to animals? The final conclusion that there are no degrees of moral status—hence, animals matter morally as much as we do—gives these questions added urgency. We will return to them in Chapter 8.

theirs—what has been called *the point of view of the universe*[1] or *the view from nowhere.*[2]

The problem here is simple: we can't do that. We can't step outside our own bodies and minds, along with the perspectives they give us, for the simple reason that we need our bodies, and our minds, to think at all. We need a perspective from which to evaluate things, and that perspective can influence what we see and what we think about what we see.

A similar challenge arises when we move from the one Mill posed (who has a better life) to the one we face: who is more important morally. We can't step completely outside of our perspective. We can't consider whether the nature of rabbits suggests they are less important from some other perspective. Granting that, we can try to be aware of the extent to which our beliefs are influenced by the way the world is versus the way it seems to us. To get a sense for how we might do this, the present chapter begins by considering how our perspective influences our views of the physical universe before turning to how it influences our views of the moral universe.

1.1. Our Place in the Physical Universe

Claudius Ptolemy was born in the Egyptian city of Alexandria in AD 100 and died there in AD 170. He was one of the more important scholars of his time and wrote on geography, music, poetry, and optics. He also refined and ultimately gave name to the belief that the planets and stars revolve around the Earth, which is the center of the universe. For thousands of years, people accepted this Ptolemaic understanding of the universe because it was consistent with everyday experience. Take a few steps and notice what happens. Things change. And the faster you move, the faster things change. The background starts to blur, and the air begins to press against your skin. A moderate walk produces a slight sensation, a brisk walk yields a gentle breeze, and 75 miles per hour on a motorcycle generates a gust of wind, to the point where you need to hold on to keep from falling off.

Now take a step outside. What do you notice? On many days at least, there is no wind. And you don't need to hold on to keep from falling down. These ordinary experiences led to the belief that the Earth is not moving. Life on a planet hurtling through space (at 67,000 miles per hour, as it turns out) would presumably be very different. We would be unable to stand against the winds. Ptolemy's picture of the physical universe thus made sense from our perspective on the world, and it explained our everyday experiences of it. His theory also made sense of what scientists saw when they looked into the sky. The sun moves across the sky from east to west. The moon rises and falls. Comets dart across the sky, like fireworks in the air.

The belief that the Earth is the stationary center of the universe is also consistent with deeper and more fundamental beliefs. It implies that *we* live at the center of the universe. The claim that everything revolves around the Earth implies that everything revolves around us. This, it was thought, is not a coincidence. We live at the center of the universe because we are its most important inhabitants: our place in the physical universe reflects our place in the moral universe.

Ptolemy's theory fit nicely with common experience until scientists started to look more closely. When they did, they saw things that did not fit with the picture Ptolemy had drawn, with a stationary Earth at the center of it all. Initially, though, these findings did not lead scientists to doubt Ptolemy's theory. Instead, they regarded them as puzzles to be solved, and, in response, scientists proposed ever more complicated theories to capture the facts they were discovering. Courtesy of these efforts, the view that the Earth is the stationary center of the universe was endorsed for another 1,400 years.

Things started to change in 1543 when Copernicus published his theory that the Earth revolves around the sun. The Copernican system has several advantages over the Ptolemaic one, including its ability to explain the seasons (as a result of the Earth being tilted relative to the sun). The invention of the telescope in 1609 allowed Galileo to observe the moons of Jupiter and propose that *they* move around Jupiter, not around the Earth. Ultimately, the view that puts the Earth on par with the other planets, and regards our solar system as just one corner of a vast universe, won the day. It provided a more complete picture of the physical universe, even though it continues to be inconsistent with the everyday experiences and the ordinary intuitions of just about everyone.

1.2. Our Place in the Moral Universe

Ptolemy was concerned with the structure of the physical universe and whether we occupy its center. We are concerned with the structure of the moral universe in the sense of determining which individuals have moral status and whether there are degrees of moral status among those who matter morally.[3] We started with the fact that the moral universe is much larger than people used to think. For much of human history, it was commonly thought that only some human beings matter morally. In many cultures, children, women, the poor, people of different religions, and those who were born in different places were regarded as having lower or no moral status. Before considering how all of this started to change, a word of caution.

The present discussion of people's beliefs about the moral universe, and the previous one regarding people's beliefs about the physical universe, come with an important qualifier: not everyone agrees. Even before Ptolemy, some people thought the Earth was not the center of the universe and, today, some people still believe it is. Similarly, with the moral universe, there have always been people who recognized that women, the poor, and people with different skin color and different hometowns have equal moral status. And, today, there are people who think they don't. These exceptions are consistent though with the fact that beliefs regarding the physical and moral universe have changed over time. The percentage of people who recognize that the Earth is not at the center of the universe has increased, and the percentage of people who recognize that women, the poor, and people with different skin color and different hometowns have equal moral status has also increased.

In contrast, and for a long time, many people continued to believe that animals are not members of the moral universe.[4] Inflicting pain on a human being they thought is a matter of moral concern; inflicting pain on nonhuman animals is not. This view was supported by a range of scientific theories, from Aristotle's belief that animals do not have rational souls, to Descartes' view that animals are just sophisticated machines. The noises animals make when they are struck or cut are analogous to the noises that emanate from the blender when you drop the spatula. It doesn't sound good, but it doesn't matter morally (unless you don't own the blender).

Immanuel Kant, one of the great philosophers of all time, developed a sophisticated theory to support this view. Kant noted that humans and animals both act on their desires. Both move toward water when they are thirsty; both sleep when they are tired. But human beings have a very different relationship to their desires. We are able to evaluate our desires and assess whether they provide good reason for us to act in a particular way. We do not simply live according to the desires that happen to be instilled in us; we consider whether the desires we happen to have are the desires we *ought* to have and whether, morally, we ought to act on or suppress them. The rabbit and I might both want to eat what's in front of us. But, unlike the rabbit, I can ask whether I should. Is it mine? Does someone else need it more? Kant thought this capacity to evaluate our own desires "raises humans infinitely above all the other beings on Earth."[5]

The belief that animals do not matter morally began to lose favor as we learned more about them. We learned that animals have experiences; they communicate with one another and learn, and they care for their offspring. Domesticating animals and taking them into our lives as pets played an important role in this process.[6] So, too, did the theory of evolution, which suggests that we are not as distinct from animals as we once thought; we share a common evolutionary history with them, we are on the same tree of life, and we share many genes and characteristics with them.[7] In particular, many animals have a central nervous system, much like ours. In us, this system gives rise to consciousness, feelings, and sensations, providing further reason to think that animals, at least those with a central nervous system, have experiences too.

These insights led to the now common belief that at least many animals are sentient, there is something that it is like to be them. Some things are good for them and others are bad for them. At a minimum, they experience pain. Those noises emanating from animals when they are stuck with a needle are very different from the noises emanating from the blender; they are not just the result of pieces of bone rubbing against one another. They indicate that the animal is experiencing pain, and it is suffering. The view that animals can suffer has led to a further expansion in our views of the moral universe. It turns out that human beings are not its only inhabitants; animals are as well. It is morally problematic to cause an animal to suffer and morally wrong unless one has good reason to do so.

Consistent with the caution mentioned earlier, some people recognized long ago that animals matter morally, while others still don't recognize that they do. This is one of the more interesting, and puzzling, aspects of debates over moral status and degrees of moral status. Thoughtful, caring, decent people can see the same things, have the same experiences, and draw completely different conclusions. This reveals the extent to which there is not a single human perspective on these issues, but many.

To briefly consider just one example, the English sport of riding to hounds involves riders on horseback following a pack of dogs through the fields, over fences and streams, with the goal of being part of the chase and ultimately witnessing the dogs catch and kill the fox or rabbit. In 1785, the English poet William Cowper described riding to hounds as the "Detested sport, that owes its pleasures to another's pain, that feeds upon the sobs and dying shrieks of harmless nature."[8] That is how some people saw riding to hounds 250 years ago. But, for others, especially the educated and cultured, mostly men, in England, it represented the height of civilization. In 1865, the English writer Anthony Trollope described watching the dogs catch and kill the fox as the "keenest pleasure." It is, Trollope wrote, "a good thing," and some of the more disappointing moments in his life occurred when the hounds failed to catch and kill the fox, or he arrived too late to witness it.[9] Over time, riding to hounds became increasingly controversial and, in 2004, hunting mammals with dogs was banned in England. But there were, and still are, dissenters. Prince Charles, heir to the British throne, now King Charles III, opposed the ban and apparently regarded it as a reason to leave his country.[10]

Current recognition that animals have moral status is analogous to the point at which it was recognized that there are other planets in the universe. As we have seen, commentators were able to accommodate that fact, but retain an elevated position for human beings, by placing the Earth at the center. It took another 1,400 years for that belief to fall out of favor. This is how most people now view the *moral* universe. Although we are not its sole inhabitants, we inhabit its center; we are morally more important, and we have greater moral status, than its other inhabitants.

Rather than offer arguments for or against this view, proponents and opponents alike tend to appeal to their intuitions in specific cases, like *Experimentation* and *House on Fire*.[11] A notable exception (we will consider a few others later on) is Peter Singer's argument in *Animal Liberation*, the book which is widely credited with initiating the animal rights movement. Human beings, Singer points out, are very different from each other. Some have IQs of 50; others have IQs of 150. Some can take care of themselves; others need to be taken care of. If the level of our moral status depended on whether we possess superior cognitive capacities, human beings with significantly lower cognitive capacities, infants and adults with severe dementia, would have lower moral status. But, Singer assumes, they don't. All human beings, he claims, have the same level of moral status. This, Singer argues, is because all humans share an important property, one that gives us moral status: we all have the capacity to suffer and to enjoy things. He then points out that many animals have this capacity, too, and morality demands that we give equal consideration to their interests in not suffering:

> No matter what the nature of the being, the principle of equality requires that its suffering be counted equally with the like suffering—insofar as rough comparisons can be made—of any other being.[12]

Proponents of degrees of moral status tend to agree that suffering matters morally, no matter whose suffering it is. Whether the individual is a baby rabbit or a baby human, the fact that it can suffer matters morally. But they deny that animals' suffering matters as much as our suffering. The philosopher Shelly Kagan writes:

> A world in which a person's toothache has been relieved is a better world than one in which the same has been done for a mouse instead.[13]

Kagan then explains why he thinks this:

> My view is that people have a higher status than animals (and some animals have a higher status than others), and that the weight we should give to a given interest—the way we should count it in our moral deliberations—does indeed depend in part on the status of the individual whose interest it is.[14]

To support the claim that we are morally more important than animals, Kagan cites our ability to imagine the past and the future, our self-awareness, our rich imaginations, our capacity to reflect on what is right and what is wrong, and our capacity to direct the course of our lives based on how we want them to go. Sentient animals either do not possess these capacities, or they possess them to lesser extents, suggesting that we are more important morally than they are.

The important point for now is this: The fact that sentient beings can all suffer does not necessarily imply that all sentient beings have the same level of moral status. It might be that all beings who are sentient have moral status and, in addition, some sentient beings, like us, have higher moral status because we possess certain properties, such as the ones Kagan mentions. To resolve the debate over whether there are degrees of moral status, then, we will need to go beyond our intuitions, and beyond the fact, emphasized by Singer, that animals are sentient, to consider whether some sentient beings matter more than others.

1.3. Moral Status and Degrees of Moral Status

To assess whether there are degrees of moral status, we first need to understand them. And, to do that, we need to understand how a world in which there are degrees of moral status differs from a world in which there aren't, a world in which everyone with moral status has it to the same degree. Imagine, then, you are going away for the weekend and you can take your cat or your dog, but not both. Most people I know take the dog, thus treating the dog better than the cat, protecting the dog rather than the cat from the anxiety of being left behind. But people who do this don't think dogs are morally more important. They take the dog because dogs get more attached to us than cats, and they experience greater distress when they are left behind. In this case, treating some individuals better than others is explained, not by a difference in degrees of moral status, but by the general principle that (causing) less distress and suffering is better than (causing) more.

Compare this case to how people used to treat kings. In most situations, it was thought important to treat the king better than everyone else. The king should eat first; the king deserves greater respect; stabbing the king is worse than stabbing a commoner. But this was not because kings are always hungrier or more sensitive to pain. Instead, it was assumed that the king should be treated better because the king is more important than everyone else.

These examples reveal that there are at least two ways in which the properties individuals possess can be morally relevant to how they ought to be treated. The first way, illustrated by the example of cats and dogs, involves individuals' properties influencing what they have at stake in a given situation. The fact that dogs get more attached to us increases how much they have at stake when they are behind for the weekend. The second way, illustrated by the king, involves individuals' properties influencing how important they are. This second way is analogous to the belief that there are degrees of moral status. The view that we have higher or full moral status regards us, in a sense, as the royalty of the moral universe.

Some proponents understand greater moral status as a kind of multiplier on the moral importance of our interests (rights or claims). On this approach, how bad it is to cause an individual to experience pain depends on how bad the pain is for the individual multiplied by the level of their moral status. If sentient animals have baseline moral status, how bad it is to cause them some pain is just a function of how bad the pain is for them. In contrast, if human beings are 1,000 times more important morally than animals, it is 1,000 times worse to cause us the same amount of pain. In other words, the same amount of pain is 1,000 times worse when it occurs in human beings than when it occurs in rabbits. This reveals that, if there are degrees of moral status, we cannot determine how we should treat others simply by looking at what they have a stake in the circumstances; we also need to know the level of their moral status.

Here is another way to think about this difference. How much each individual has at stake, and who has more at stake, tends to vary from situation to situation. Hence, who should be treated better will vary. The fact that someone has claustrophobia provides a reason to protect them when it comes to entering a narrow mine shaft to save a trapped child. But it does not provide a reason to treat them better when it comes to deciding who goes on a turbulent boat ride. In that case, we treat the individual who experiences motion sickness better. Having greater moral status, in contrast, provides a reason to treat individuals better in most or all circumstances. In just about all situations, we ought to treat the king better. When deciding whom to protect from the risks of going down a narrow mine shaft or going on a turbulent boat ride, we don't ask whether the king has claustrophobia or gets motion sickness. We protect the king because they are more important; hence, it is worse to send them even when the harms they would experience are no greater.

This example explains how degrees of moral status can simplify the moral life, and how rejection of degrees of moral status can complicate it. The conclusion that there are no degrees of moral status implies that we can't determine who should be treated better simply by determining who has higher moral status. We need to consider what the situation involves, what characteristics each individual has, and how the combination of situation and characteristics influences who has more at stake. Who is hungrier? Who has claustrophobia? Who gets worse motion sickness? Because there are so many properties which can influence what individuals have at stake, the list goes on and on, revealing that making these assessments will be harder and take more time. Moreover, these comparisons often do not yield clear answers. It might not be clear, despite careful consideration, who is hungrier or who is more likely to experience motion sickness. A world without degrees of moral status is thus more complicated: it will be harder to figure out what we ought to do and, once we have decided, it will be harder to know whether we made the right choice, whether ethically we did the right thing.

The claim that the king is more important than everyone else, and should be treated better in essentially all circumstances, raised an obvious question: Why are the interests of the king more important than the interests of commoners? Why is it more important to respect the king than it is to respect everyone else? The influential British politician and theorist Edmund Burke, who lived from 1729 to 1797, gave the following answer:

> The Author of our being is the Author of our place in the order of existence—and that, having disposed and marshaled us by a divine tactic, not according to our will, but according to His, He has in and by that disposition virtually subjected us to act the part which belongs to the place assigned us.[15]

Throughout history, many people have given the same answer for why human beings are morally more important than animals: God made us more important. Contemporary proponents of degrees of moral status, as we have seen, give a different answer. They argue that we have higher moral status than animals because we possess one or more superior cognitive capacities. To evaluate this view, then, we need to assess whether any properties can have this effect: Can some properties make the interests, claims, or rights of those who possess them stronger than the analogous interests, claims, or rights of individuals who have moral status, but who lack the properties in question (or possess them to a lesser extent)?

1.4. Being More Important Versus Having More at Stake

According to proponents, the fact that we possess superior cognitive capacities implies that it is worse to harm us than it is to harm animals, even when the amount of harm is similar. It is worse to cause us a given amount of pain and suffering than it is to cause the same amount of pain and suffering to a rabbit. One might be tempted to reject this view from the start. How could possession of some property be relevant to how we ought to treat an individual when it has no impact on what the individual has at stake in the circumstances? Whether it is better to put rabbits or human beings in the pain-inducing experiment seems to depend on who will suffer more. How could the fact that I am a moral agent or self-conscious make it worse to cause me pain when it doesn't affect the magnitude of my pain or the extent of my consequent suffering?

One way to think about this question is in terms of how the same level of pain and suffering could have different moral significance depending on its location, whether it is located in a human being or a rabbit. How can the location matter morally when it doesn't affect the amount of pain and suffering? The king suggests a possible answer. Possession of some property might imply that some individuals should be treated better in most or all circumstances, not because they tend to have more at stake, but because they are more important. If the individuals themselves are morally more important, they should be treated better, even when they do not have more at stake. This explains why, as we considered in section I.3, only some properties offer plausible bases for greater moral status. The fact that humans blush does not provide a plausible basis for higher moral status because it clearly does not make us more important. Superior cognitive capacities, in contrast, are important; hence, it seems plausible to think that possession of them could make us more important. In effect, we do not simply possess the capacity for moral agency; we are moral agents.

To evaluate this view, we will need to keep in mind that possession of superior cognitive capacities influences what we have at stake in a wide range of circumstances. Shelly Kagan, as noted earlier, claims that individuals' level of moral status depends on a number of cognitive capacities, including the extent to which they are aware of the past and future. If Kagan is right, the fact that human beings have greater awareness of the past and future gives us greater moral status than rabbits and, thereby, provides a reason, in *House on Fire*, to save the human being rather than the rabbit. But notice that greater awareness of the past and future might result in the human being experiencing more suffering in *House on Fire* compared to the rabbit. In addition to the pain caused by the smoke and heat, the human might suffer significant distress thinking about the future they will lose if they die, and worrying about who will take care of their children when they are gone. The rabbit's decreased awareness of the future likely protects it from this source of distress.

This highlights a critical point: The fact that we ought to treat one individual better than another because the first individual possesses one or more superior cognitive capacities does not show that the first individual has higher moral status. To make that determination, we need to assess whether we should treat the first individual better because possession of the superior cognitive capacity gives them higher moral status or because it implies that they have more at stake in the circumstances. The importance of distinguishing between these two possibilities is illustrated by the following passage:

> I firmly believe that man is the measure of all things. If the mating habits of the Arctic caribou have to be disturbed so we can produce 1 million barrels of oil a day—on a drilling footprint the size of Dulles Airport in a refuge the size of Ireland—I say: Apologize to the amorous herd, then drill.[16]

The author believes that we are justified in disturbing the caribou's mating habits in order to realize these benefits. And the fact that this involves acting in ways that benefit us leads him to conclude that we must have higher moral status (we are the "measure of all things"). But the distinction between individuals' level of moral status and what they have at stake reveals that this does not necessarily follow. It might be that humans should be treated better in this case, not because we have higher moral status, but because we have more at stake. It might be that the benefits to us from so much oil significantly outweigh the harms to the Caribou. After all, in a refuge the size of Ireland, there must be other places they can go.

To show that we have higher moral status, we need to show that we should be privileged *even when we do not have more at stake. House on Fire* and *Experimentation* are designed with this challenge in mind. In both scenarios, the rabbits and the human beings have the same amount at stake. Each will suffer to the same degree in *Experimentation*; each will die in *House on Fire*. If everything else is equal in these cases (later on, we will consider whether they are), the intuition that we should treat the human being better cannot be explained by a difference in what they have at stake compared to the rabbit; hence, it must be explained by their having higher moral status. This is the claim we will assess.

1.5. Three Morally Significant Harms and Wrongs

There are, unfortunately, many ways to treat others badly and wrongly. And a plausible view maintains that the increased protections enjoyed by individuals who have higher moral status are stronger with respect to every, or essentially every, one of them. With that said, it will not be possible to consider all the different possible harms and wrongs and assess whether it is always worse to inflict them on humans than rabbits. Instead, taking a cue from *Experimentation* and *House on Fire*, we will consider three of the more important ways to treat others negatively: (1) killing them; (2) causing them to suffer; and (3) failing to respect them.

It seems clearly and significantly worse to kill a human being than it is to kill a rabbit. This is not to say that there are no moral concerns in the latter case. Imagine someone raises rabbits and kills them for no reason at all. That is morally problematic. In contrast, raising rabbits to have a good life and then killing them painlessly as part of an important medical experiment might be acceptable (section 8.5 considers whether it is). This suggests there is moral resistance to killing rabbits, but it can be outweighed by important benefits to us. The morality of killing human beings is very different. It is not simply problematic to intentionally kill human beings as part of medical research; it is unethical. This supports the claim that there are degrees of moral status.

Second, it seems worse to cause human beings to have aversive experiences and to suffer. As we will consider in section 8.5, this belief is reflected in almost all research guidelines, which stipulate that testing with animals should precede testing with human beings. The Nuremberg Code, perhaps the most important statement on research ethics ever written, maintains that experiments with human beings are ethical only when they are "based on the results of animal experimentation."[17] This requirement reflects the intuition that, even when the level of suffering that subjects experience will be the same, it is worse to cause human beings to suffer than it is to cause animals to suffer. This intuition provides further support for the twin claims that there are degrees of moral status, and we are morally more important than animals.

Third, respect for individuals as individuals and respect for individuals' agency seem relevant to human beings, but not animals. Perhaps the most famous articulation of respect for individuals as individuals traces to Immanuel Kant, who argued that this sense of respect requires that individuals are treated as *ends in themselves*, never as mere means for accomplishing the goals of others. In a classic example, respect for individuals as individuals implies that it is unethical to kill one human being in order to use their organs to save five others. In contrast, it is widely thought to be acceptable to sacrifice rabbits to benefit human beings, suggesting that respect for individuals as individuals does not apply to rabbits.

Respect for individuals' agency, understood broadly, refers to allowing individuals to live their lives as they attempt to live them. Failures of respect in this sense involve disrupting or preventing others from doing what they want to do. Intuitively, this sense of respect seems extremely important when it comes to human beings. In contrast, disrupting the lives of rabbits seems to raise little ethical concern. Imagine your friends getting together and deciding that they will choose who you marry, or whether you marry at all. This would be seriously problematic, even if things turn out for the better. We might be better off, we might be happier and more content with our lives, if our friends and families chose our partners for us. Nonetheless, others should not interfere in our lives in this way. In contrast, it does not seem problematic for us to decide who rabbits spend their lives with, whether they have opportunities to mate and when. This, too, seems to suggest that we are morally more important than rabbits.

Proponents of degrees of moral status are not necessarily committed to our being morally more important than animals in all three ways. Some proponents might think that our greater moral status affords us greater protections against being killed and treated with disrespect, but not against being caused to suffer.[18] Or they might think that it is equally bad to kill human beings and rabbits, but respect applies only to human beings. Granting these possibilities, every endorsement of degrees of moral status maintains that it is morally worse to harm or wrong individuals with higher moral status in at least one or more significant ways. Given that these three represent some of the most significant ways to harm or wrong others, they provide a way to evaluate whether they are degrees of moral status. The question we will try to answer, then, is this: Are there properties which imply that it is worse to kill, cause to suffer, and/or disrespect those who possess them compared to individuals who have moral status but do not possess the properties in question (at least to the same extent)?

Summary

Morally, the right thing to do is often to treat individuals differently. This is frequently because the individuals are different in ways that influence what they have at stake in the circumstances. We saw this with respect to whether it is better to leave the cat or the dog behind for the weekend. People faced with this choice often leave the cat, not because they think cats are less important morally, but because the cat is less likely to be traumatized.

Some people believe that what individuals have at stake is all that matters when it comes to determining how they ought to be treated. Many others believe that there is more to morality than what individuals have at stake; there is also their level of moral status. We considered this possibility with respect to chasing the cat off the sofa and giving the spot to a human being. People who do this typically don't take the time to figure out who has more at stake, the cat or the human being. Instead, they give the spot to the human being because they assume that human beings are morally more important than cats. To understand this possibility, we considered how people used to think about kings. The king, it was assumed, should be treated better in just about every case, not because they always have more at stake, but because they are more important. The same amount of hunger, the same amount of pain, is worse when it is located in the king than when it is located in a commoner.

The claim that there are degrees of moral status gains support from our intuitions regarding three important ways of treating others negatively: killing them, causing them to suffer, and failing to respect them. In each case, it seems worse to do these things to human beings than to animals. Our challenge, then, is to investigate this possibility by assessing whether human beings possess one or more properties which explain these differences by implying that we are morally more important than animals. Chapter 2 provides some background for this assessment, while Chapter 3 provides the assessment. The conclusion that there are no such properties, hence, no degrees of moral status, leads, in Chapters 4–6, to a reassessment of whether it really is worse to kill, cause to suffer, and fail to respect human beings. Chapter 7 considers whether, for pragmatic reasons, we should believe in degrees of moral status anyway. After concluding that we shouldn't, Chapter 8 considers seven important implications for a world without degrees of moral status.

2
The Possibility of Degrees of Moral Status

Introduction

Many people have strong intuitions that human beings matter morally more than animals. These intuitions, and their potential implications, provide reason to consider systematically whether there are degrees of moral status among those who matter morally. The present chapter sets the stage for this assessment. It first considers the claim that degrees of moral status can be dismissed without the need for systematic assessment. After rejecting this claim, it describes and distinguishes three types of morally relevant properties: moral status conferring properties, moral action guiding properties, and moral status enhancing properties. Finally, it argues that degrees of moral status require moral status enhancing properties. This sets the stage, in Chapter 3, for assessing whether any properties are moral status enhancing.

2.1. Can We Reject Degrees of Moral Status Out of Hand?

How we ought to treat others depends critically on what they have at stake in the circumstances. And what they have at stake depends on the properties they possess. Some things, like grains of sand, don't possess any properties which give them a personal stake in what happens to them. Others do. They get anxious when they are

Life Without Degrees of Moral Status. David S. Wendler, Oxford University Press.
DOI: 10.1093/oso/9780197675328.003.0003

left behind for the weekend; they experience pain when they are stuck with needles; they have claustrophobia or motion sickness. Some people argue that this is the only way in which the properties individuals possess can influence how they ought to be treated. If that's right, it follows that there are no properties which increase individuals' level of moral status. The philosopher James Rachels, who endorsed this view, writes:

> There is no characteristic, or reasonably small set of characteristics, that sets some creatures apart from others as meriting respectful treatment. That is the wrong way to think about the relation between an individual's characteristics and how he or she may be treated. Instead we have an array of characteristics and an array of treatments, with each characteristic relevant to justifying some types of treatment but not others. If an individual possesses a particular characteristic (such as the ability to feel pain), then we may have a direct duty to treat it in a certain way (not to torture it), even if that same individual does not possess other characteristics (such as autonomy) that would mandate other sorts of treatment (refraining from coercion).[1]

Rachels is claiming that the moral relevance of all the properties individuals possess traces to the fact that they influence what the individuals have at stake in specific circumstances.[2] In contrast, he claims, there aren't any properties which increase the moral importance of the individuals themselves. The moral significance of a given experience of pain and suffering, in Rachels's view, depends solely on how bad it is, not on the individual in which it occurs. Proponents of degrees of moral status reject this view; they argue that the individual in which an aversive experience occurs, whether it occurs in a human being or a rabbit, can influence its moral significance. Baruch Brody, as we saw in section I.2, thinks that "the same unit of pain counts less morally if experienced by an animal than it would if experienced by a human being."[3]

Brody is clearly positing something in addition to the fact that individuals have properties which influence the extent to which different treatments cause them to experience pain and to suffer. At least implicitly, he is positing the existence of one or more properties which influence the moral status of the individuals who possess them. Rachels denies this. He claims that the nature of the pain experience itself, how bad it is for the individual who experiences it, is all that matters morally. He might be right, but we can't simply assume that he is. We can't assume that all the properties individuals possess work in the way Rachels claims. We need to find out whether they do, and for that, we need to answer the following question: Are there any properties which make the individuals who possess them morally more important which, in turn, make their interests, rights, or claims more important than the analogous interests, rights, or claims of individuals who lack the properties in question? If there are, Brody is right. If there aren't, Rachels is right.

One way to see the need to go beyond assumptions and intuitions is to notice that some properties do influence how we ought to treat individuals, without affecting what they have at stake in the circumstances. For example, when it comes to risky medical research, it is morally better to enroll individuals who can consent rather than individual who cannot consent, even when the risks and benefits to the two groups are the same. Similarly, the ethics of putting individuals in prison, whether doing so is seriously unethical or not, depends on more than just what the experience is like for them and what they have at stake. It depends on whether they committed a crime which merits imprisonment.

Examples like these reveal that the properties individuals possess, whether they can consent, whether they committed a crime, can influence how they ought to be treated, without influencing what they have at stake. In these cases, it is the actions of the individuals, whether they consented, whether they committed a crime, which influence how they ought to be treated. To evaluate the possibility of degrees of moral status, then, we need to consider whether there are other properties which can influence how we ought to treat others in a different way, based not on what they have done, but based on who they are, on how important they are morally.

2.2. Three Types of Morally Relevant Properties

To determine whether there are any properties which can increase the moral status of those who possess them, it will be helpful to distinguish three types of morally relevant properties (for more, see Box 2.1: A Catalog of Morally Relevant Properties). First, there are properties which endow their possessors with moral status, what I will call moral status *conferring* properties. Previously, I suggested that sentience is a moral status conferring property: individuals who are sentient have moral status; they matter morally (we will consider this view in more depth in section 6.3). Albert Schweitzer believed that being alive is moral status conferring, that ethics involves maintaining, assisting, and enhancing "all that lives."[4] Still others believe that being an agent or a moral agent is moral status conferring (section 6.2 discusses these views).

The fact that an individual possesses one or more moral status *conferring* properties implies that they are a member of the moral community, that how they are treated matters morally. But the fact that an individual has moral status does not reveal *how* they ought to be treated. It does not reveal whether it is better to give them a hand or stick them with a needle. To make that determination, we need to look to a second type of property, what I will call *moral action guiding* properties. The capacity to experience pain and to suffer is an example. Possession of this property implies that certain ways of treating an individual are morally problematic, namely, ways that cause them to experience pain and to suffer. These are

Box 2.1 A Catalog of Morally Relevant Properties

Moral status involves an individual mattering morally for their own sake. Depending on one's preferred theory, this means that their interests are morally relevant, they have rights, they merit respect, and/or they make claims on others regarding how they ought to be treated.

Moral status conferring properties endow individuals with moral status; they imply that the individuals matter morally for their own sakes. Properties commonly cited as moral status conferring include being alive, sentience, agency, and moral agency.

Moral action guiding properties are relevant to the ethics of specific ways of treating individuals. The fact that some animals are made anxious by sounds at high frequency is one example. It is relevant to the ethics of exposing them to these sounds for an extended period of time. In contrast, possession of this property is not relevant to the ethics of confining the animals in small cages for an extended period of time.

Moral status enhancing properties increase the moral status of the individuals who possess them. Individuals who possess one or more moral status enhancing properties matter more morally than individuals who are otherwise identical but lack such properties. Depending on one's preferred theory, this might mean that their interests have greater moral weight, they have rights, their rights are stronger, they merit respect, they merit greater respect, and/or they make stronger claims on others regarding how they ought to be treated.

Moral status decreasing properties diminish the moral status of the individuals who possess them. Individuals who possess one or more moral status decreasing properties matter less morally than individuals who are otherwise identical but lack such properties. Proponents of degrees of moral status typically do not endorse moral status decreasing properties. However, in Chapter 3, we will consider whether endorsement of degrees of moral status commits proponents to the possibility of moral status decreasing properties, and whether the thresholds needed for degrees of moral status require moral status decreasing properties.

the properties Rachels emphasizes. They are, he claims, the only properties which are relevant to how individuals ought to be treated.

As we have seen, proponents of degrees of moral status disagree; they assume that one or more properties increase individuals' moral status. For example, many proponents regard possession of one or more superior cognitive capacities as making individuals themselves morally more important. I will describe this view in terms of the claim that the cited capacities are moral status *enhancing* in the sense that they increase individuals' moral status. Again, depending on one's preferred moral theory, this implies that the individuals' interests, rights, or claims are stronger than the analogous interests, rights, or claims of individuals who have moral status but lack these properties.

A given property can belong to one or more of these three categories. Proponents of degrees of moral status regard our superior cognitive capacities as increasing our moral status; to that extent they are moral status enhancing. These properties also influence what individuals have at stake in many situations and are thereby moral action guiding. To take another example, I will argue in section 6.3 that sentience endows individuals with moral status. To that extent, sentience is moral status conferring. But sentience is also an important moral action guiding property. The fact that an individual is sentient gives us reason not to treat them in ways that lead to their having aversive experiences, to their experiencing pain and suffering, fear and anxiety.

2.3. Why Degrees of Moral Status Require Moral Status Enhancing Properties

Section 2.2 described three types of morally relevant properties: moral status conferring properties, moral action guiding properties, and moral status enhancing properties. The fact that some individuals have moral status, while others don't, reveals that one or more properties must be moral status conferring. This raises the present question of whether some individuals with moral status are more important morally than others.

We know that moral action guiding properties, such as the capacity to feel pain and to suffer, long legs, weak stomachs, and claustrophobia, exist. The most straightforward way to show that there are degrees of moral status, then, would be to argue that possession of moral action guiding properties can increase individuals' level of moral status. To assess this possibility, it will help to briefly consider the nature of moral action guiding properties.

One of the more important aspects of moral action guiding properties is that they are relevant to some ways of treating individuals, but not others. Consider the capacity to feel pain. The fact that rabbits can experience pain is relevant to the ethics of experiments which involve sticking them with needles, but it is not relevant to the ethics of experiments which involve killing them painlessly. Similarly, the fact that some animals are sensitive to high-frequency sounds is relevant to the ethics of experiments which expose them to such sounds for extended periods. This property is not relevant to the ethics of experiments which involve confining the animals in small cages for extended periods. Put generally, to determine whether there are moral concerns with a particular way of treating some animals, one needs to know whether the animals have any capacities by which the treatment (sticking them with needles, confining them in small cages) might benefit or harm them, whether the treatments might make their lives go better or worse for them. Alternatively, is there some way in which the treatment fails to respect them appropriately?

The fact that moral action guiding properties gain their moral relevance in virtue of being a link in specific causal pathways that ultimately affect the individuals who possess them explains why they are relevant to some treatments, but not others. It also points to another feature of moral action guiding properties: their influence typically occurs prior to the harm or wrong in question. The nerve fibers that underlie the capacity to feel pain in human beings connect certain actions (e.g., stabbing) to the experience of pain. Eliminating the fibers eliminates the moral concern by eliminating the outcome. Of course, stabbing individuals might implicate other moral action guiding properties. For example, individuals with congenital insensitivity to pain don't feel physical pain. But stabbing them can harm them in other ways, by causing tissue damage, bleeding, or infection. These examples provide insight into how we can determine whether a given action is moral action guiding.

Sharks have small pores, called ampullae of Lorenzini, around their heads which allow them to sense electrical signals coming from the muscle movements of other animals.[5] This is an extremely valuable sense for a predator that often lives in dark oceans. To determine whether this capacity is moral action guiding, we would assess whether, in addition to conveying information about the world, it also conveys or leads to harms or wrongs for the sharks. For example, robots also rely on signals to navigate the environment. However, this capacity is not morally relevant because at least current robots are not sentient; hence, it does not have the potential to make things go better or worse for the sake of the robot.

Our question, then, is whether sharks are aware of the signals conveyed by their ampullae of Lorenzini. Do they become anxious, for example, when they sense these fields? Are these pathways connected to pain or pleasure centers in the shark brain? Alternatively, the ampullae of Lorenzini might allow sharks to carry out plans or projects that are morally significant. Perhaps they allow the sharks to find mates and reproduce and, therefore, have families and carry on the species. And it might be good for individual sharks to do this because having a family and helping to maintain the species leads to a better life for the individual shark. If it does, it would follow that the capacities sharks have as a result of possessing the ampullae of Lorenzini are moral action guiding properties.

The connection between moral action guiding properties and specific situations or ways of treating individuals suggests that they do not render individuals morally more important across a range of otherwise independent contexts or treatments. It follows that individual moral action guiding properties cannot increase the moral status of those who possess them. This conclusion raises the question of whether individuals' level of moral status might be influenced by *how many* moral action guiding properties they possess. If so, differences in degrees of moral status would be possible without moral status enhancing properties. And, since we know that moral action guiding properties exist, this would make degrees of moral status no more mysterious or speculative than moral status itself.

Perhaps the most straightforward version of this view maintains that an individual's level of moral status is determined by the number of moral action guiding properties they possess. Assuming that human beings possess a significantly greater number of moral action guiding properties than animals, this view would provide a basis for our greater moral status. Some philosophers believe that there is essentially only one moral action guiding property: the capacity to suffer. If that is right, there are no differences in the number of moral action guiding properties which could provide a basis for degrees of moral status. Even if we add to this the capacity for moral agency (or autonomous action), humans would have just one more moral action guiding property than rabbits, which does not seem enough of a numerical difference to generate a difference in degrees of moral status.

To address this concern, proponents would need to argue that this is not the right way to count moral action guiding properties: we do not have one or two; we have many. To that end, they would need to identify the right way to count moral action guiding properties. Is the capacity to suffer one moral action guiding property? Or do the different ways of suffering count as distinct moral action guiding properties: the capacity to suffer as the result of pain, as the result of anxiety, as the result of loneliness, in which case our capacity to suffer might represent forty or fifty moral action guiding properties. Although each of these options might represent hundreds of different moral action guiding properties? The capacity to feel pain as the result of needle sticks, as a result of stubbing one's toes, and so on. The fact that the answers to these questions are not clear undermines this possibility for establishing degrees of moral status.

This account would also make it possible to increase individuals' level of moral status in odd ways. If I promise to do something for you, you gain a morally relevant claim that I do it. More promises lead to more morally relevant claims. On the present account, then, we could give an individual higher moral status simply by making them many promises. But the fact that an individual is owed a favor, even the fact that they are owed five thousand favors, is irrelevant to the moral significance of enrolling them in experiments which involve their being stuck with needles and, hence, to the level of their moral status.

Proponents might respond that individuals' level of moral status is not determined simply by how many moral action guiding properties they possess. It is determined by how many *significant or important* moral action guiding properties they possess. On this view, it doesn't matter how we count the number of claims an individual gains when we promise to take them to dinner, nor does it matter how many times we make such a promise. What matters is how many *fundamental* moral action guiding properties they possess. With this in mind, Mary Ann Warren endorses a multicriteria account according to which individuals' level of moral status is determined by how many of the following properties they possess: being alive, sentience, moral agency, being human, being part of an ecological system, and being a member of a social community.[6]

One problem with this approach is that the importance of a given property is not determined solely by the nature of the property. It depends on the role that the property plays in the life of an individual or group of individuals. For example, article 10 of the Universal Declaration of Human Rights states: "Everyone is entitled in full equality to a fair and public hearing by an independent and impartial tribunal, in the determination of his rights and obligations and of any criminal charge against him."[7] The right to respond to criminal charges is possessed by human beings, but not rabbits. But, this right is not important for rabbits; hence, the fact that rabbits lack it does not provide reason to think that they are morally less important than we are.

This analysis suggests that moral status cannot be developed as it were from the bottom up. Individuals cannot attain higher moral status simply by acquiring more moral action guiding properties. Views which endorse only one or two moral action guiding properties cannot generate sufficient differences to yield differences in levels of moral status. Even if we posit a greater number of moral action guiding properties, there is no clear way to count them. And, as we will see in section 3.5, even if there were, the present view would be undermined by the absence of morally relevant thresholds. Finally, whether a given property is important depends on the individual in question; hence, failure to possess the property does not necessarily imply that an individual is less important morally.

A different way to support this approach would be to show that animals lack certain rights which are critical for their welfare or well-being. For example, a finding that human beings, but not rabbits, have a right against being tortured would suggest that we are more important morally than rabbits. But if rabbits lack this right, it presumably is because they have lower moral status than we do. In other words, the lack of this right would not make animals less important morally. It would be a result of the fact that they are less important morally and that fact would need to be explained by something else, by the fact that they lack one or more moral status enhancing properties which we possess.

Summary

Some people, like James Rachels, maintain that how we ought to treat others depends solely on what they have at stake in the circumstances. If that is right, there are no degrees of moral status. To begin to assess whether it is, we considered three types of morally relevant properties: moral status conferring properties, moral action guiding properties, and moral status enhancing properties. Moral status conferring properties determine whether individuals have moral status at all. Moral action guiding properties determine how individuals with moral status ought to be treated but, as we have seen, do not increase their level of moral status. This suggests that the existence of degrees of moral status depends on moral status enhancing properties. It depends on the existence of some properties which increase the moral status of those who possess them compared to other individuals who have moral status but lack the property or properties in question.

Rachels's rejection of degrees of moral status amounts to the claim that there are no moral status enhancing properties. He may be right and, if he is, there are no degrees of moral status. But we can't simply assume that he is. We need to see whether moral action guiding properties are all there is to how we should treat others, or whether some individuals also possess moral status enhancing properties. Answering that question will determine whether there are degrees of moral status. That is the challenge for Chapter 3.

3

Are Some Properties Moral Status Enhancing?

Introduction

Some philosophers think that animals lack moral status because they lack one or more superior cognitive capacities, such as the capacity for moral agency. This view, which regards certain superior cognitive capacities as moral status *conferring*, is refuted by the fact that rabbits have moral status, even though they lack the cited superior cognitive capacities. Proponents of degrees of moral status accept this expansion of the moral universe and attempt to retain a privileged position for human beings by maintaining that the cited superior cognitive capacities are moral status enhancing rather than moral status conferring. These properties do not determine whether an individual matters morally at all; they increase the moral status of the individuals who possess them. For example, many proponents of degrees of moral status regard sentience as moral status conferring (it endows individuals with baseline moral status) and the cited certain superior cognitive capacities as moral status enhancing (they endow us with higher or full moral status). This view explains why the pain and suffering of rabbits matters morally, while still capturing the intuition that we matter more than rabbits.

The present challenge is to assess this possibility by assessing whether there are any properties which are moral status enhancing. The question here is not whether the superior cognitive capacities that proponents of degrees of moral status cite in fact exist. We are

Life Without Degrees of Moral Status. David S. Wendler, Oxford University Press.
DOI: 10.1093/oso/9780197675328.003.0004

not going to question whether some individuals are moral agents, or autonomous, or self-conscious, or have a sense of the past and the future. Some individuals do; others don't. The question is whether one or more of these properties, undoubtedly important as moral action guiding properties, are also moral status enhancing. Does possession of one or more of these properties increase the moral status of those who possess them?

In the present chapter, we will discuss five considerations which reveal that there are no such properties. This conclusion, combined with the previous conclusion that degrees of moral status require moral status enhancing properties, implies that there are no degrees of moral status. Once this conclusion is in place, the remainder of the book considers its implications for rabbits, robots, and the rest of us.

3.1. An Excess of Candidates

Different proponents cite different properties as moral status enhancing, including being alive, being sentient, being part of an ecological system, having intimate relationships, being a member of a social community, being human, having self-consciousness, being an agent, having the capacity for moral agency, having moral personhood, having the capacity to imagine the past and the future, having rich imaginations, being autonomous, and having narrative self-identity. As long as we assume that human beings have higher moral status than animals, this disagreement is not terribly important. We can assume that our higher moral status traces to one or more of the properties that we possess and which animals don't possess (at least to the same extent). But once we question whether there are degrees of moral status at all, this disagreement, and the length of the list, becomes important. It raises the need to consider whether any properties might be moral status enhancing and, if so, which ones.

Previously, I noted that humans blush, but animals don't. This difference does not provide a basis for differences in degrees of moral status because it is not itself important enough to make some individuals morally more important than others. With this in mind, proponents of degrees of moral status tend to cite especially impressive or important properties. Moral agency, the capacity to recognize moral principles and act according to them so impressed Immanuel Kant that he regarded it as the one and only moral status *conferring* property: it determines whether individuals have moral status at all.[1] Proponents might try to use a similar argument to show that moral agency is moral status *enhancing*. It is so impressive that it makes us morally more important and, thereby, endows us with greater moral status than animals.[2]

The first problem with this approach is that other properties are impressive as well, including the many other properties that proponents cite as moral status enhancing, including self-consciousness, a narrative self-identity, a sense for the past and future, and moral personhood. Each of these properties is impressive—their advocates would not have cited them if they weren't. If it is simply the level of impressiveness that makes a property moral status enhancing, it seems to follow that these and countless other properties qualify. Consider the capacities to love, to create art, and to reproduce. They are impressive. Moreover, it is not clear whether they are more or less impressive than moral agency or autonomy. We are thus left with far too many moral status enhancing properties, including ones possessed by animals.

Proponents might respond that it is not just the significance or the impressiveness of our superior cognitive capacities which makes them moral status enhancing. It is something more. Appealing to Kant again suggests one possibility: transcendence. As Kant notes, the capacity for moral agency is impressive in a particular way. It involves our transcending the limitations placed on other animals by their desires and inclinations. They simply act based on the desires and inclinations they happen to have without the capacity to consider whether the way they are inclined to act—attack a deer, say—is the way that morally they ought to act. Our ability to transcend our inclinations and desires, to ask whether we ought to act in the way we are inclined to act, involves an impressive kind of transcendence.

Granting that transcendence is impressive, it is not clear why transcendence might provide the basis for increased moral status as opposed to other ways of being impressive, such as the capacity to create art. Why should we think that one is more important than the other? Moreover, there are different types of transcendence. The primary reason why the capacity to fly seems so impressive, for example, is that it involves the capacity to transcend gravity to soar the skies. Or, to introduce a character we will consider in detail in section 4.2, one that Kant did not know about, consider the immortal jellyfish. One of the most significant forces or natural laws is mortality. Death seems to be an inevitable, inescapable fact of life. If you are alive, you will die, and there is nothing you or anyone else can do about it. But what if there were beings that transcend mortality, potentially living forever? The immortal jellyfish might be one. If importance and transcendence together account for the fact that moral agency endows us with greater moral status, it seems to follow that the immortal jellyfish has high moral status as well. Similarly, the capacity to reproduce involves transcending our own mortality in a different way, while the capacity to love involves transcending ourselves. So, too, does the capacity to act selflessly in the sense of helping others and doing things for them, at some risk or cost to oneself. Once again, we are left with far too many moral status enhancing properties.

Proponents might respond that, in addition to being important and transcendent, moral agency is different from these other properties because it is relevant to morality. Several challenges arise here. Many of the other capacities we have considered are relevant to morality as well. The capacity for moral agency involves going beyond one's desires to do what is moral. Acting selflessly involves going beyond what is good for oneself to benefit others. Consider the wonder of caring as much or more for your partner or child than for yourself. Or the morally incredible feats of soldiers who sacrifice themselves to save their fellows. Or the more common, but still amazing, choice to donate a kidney to a loved one, or even to a stranger. Clearly the capacity to act selflessly is relevant to morality, and it seems to have great moral significance. And those who possess it deserve great respect. At the same time, it doesn't increase the level of their moral status. If it did, sentient animals that are significantly more selfless than humans might be more important morally than we are. The reason why they aren't is that selflessness is important, even transcendent, and it is morally relevant. But it doesn't make individuals morally more important.

Similarly, immortal jellyfish, which may be able to transcend mortality, are not more important morally than Earth-bound mortals. If impressive, transcendent, and morally relevant properties were moral status enhancing, these individuals would have higher moral status. Finally, countless important properties are relevant to morality in other ways. Consider the capacity to create art. It can be used to challenge the mistaken beliefs of others, depict others in morally problematic ways, and to raise money for charitable causes. Still, these properties do not increase individuals' moral status, suggesting that no properties, even properties which are impressive, transcendent, and morally relevant, do. The remaining sections of the present chapter explore and ultimately support this conclusion.

3.2. The Improbability of Moral Status Decreasing Properties

Postulation of degrees of moral status assumes that some properties *increase* individuals' moral status. However, that is possible only if some properties can *influence* the level of individuals' moral status.

But, if that is possible, other properties might *diminish* the level of individuals' moral status. In other words, a world in which moral status enhancing properties exist would also be a world in which moral status decreasing properties might exist. In response, Shelly Kagan, a proponent of degrees of moral status, writes:

> I simply find myself unable to identify any psychological capacities—or for that matter, any other features at all—that strike me as being negative in the relevant sense, features where I find myself inclined to judge that if only the individual lacked the relevant capacity altogether, that would directly result in a higher moral status.[3]

I agree with Kagan that there are no properties which diminish the moral status of the individuals who possess them. But the critical question for our discussion is: why not?

Human beings possess many negative capacities: the capacity for jealousy, for intentionally inflicting suffering on others, for being indifferent to the suffering of others, for genocide. According to a prominent historian, the primary lesson of the twentieth century is that human beings have the capacity to "Treat other human beings as members of an inferior and indeed malignant species—as mere Vermin."[4] Granting this unfortunate fact about human beings, it does seem, as Kagan claims, that the negative capacities we possess do not lower our moral status; eliminating these capacities would not result in our having higher moral status. Why is that?

If the capacities we possess could influence our level of moral status, these capacities would be clear candidates; they seem as negative as the capacity for moral agency is positive. The best explanation, then, for why our negative capacities do not diminish our level of moral status is that our capacities cannot influence the level of our moral status. And, if that is right, the capacity for moral agency or other superior cognitive capacities can't influence our level of moral status either. That is, the fact that there are no properties which can diminish our moral status suggests that there are no properties which can enhance our moral status. This supports the conclusion from section 3.1 that there are no moral status enhancing properties.

The fact that human beings have the capacity to do terrible things poses another challenge for proponents of degrees of moral status. They argue that we are morally more important than rabbits because we possess certain superior cognitive capacities, such as the capacity for moral agency or autonomy. And these capacities increase our moral status because they are especially impressive or valuable. To this point, we have accepted the assumption that these capacities are positive or valuable. However, the present discussion highlights an important point: the very same capacities that proponents of degrees of moral status cite enable us to do terrible things. For example, Kagan argues that our higher moral status traces, in part, to our capacity to imagine the past and the future. This capacity enables us to do great things, to learn from the past and do better in the future. Yet this same capacity also enables us to plan and carry out slaughter and genocide. Similarly, while our capacity to recognize that other people have minds endows us with the ability to treat them with respect, it also endows us with the ability to torture them.

If these capacities provide the basis for both the wonderful and the horrible things humans do and have done, it raises the question of how we determine whether they qualify as positive or negative in the first place. Or perhaps it reveals that our superior cognitive capacities are not simply positive or negative; they are both. They enable us to do wonderful and horrific things. It follows that, even if our capacities could influence the level of our moral status, possession of the capacities cited by proponents of degrees of moral status could increase and decrease it. Depending on how the ultimate balance works out, possession of these capacities might result in our having lower moral status than animals; it might even turn out that we have negative moral status overall.

Proponents of degrees of moral status might object that moral status is a kind of protection for individuals. It says that they matter morally for their own sake, that it is problematic to harm them and important to respect them. On that understanding, it is not possible for anyone to have negative moral status. However, there is another way to think about moral status, not as directing us to avoid harming individuals, but as a marker for how they morally ought to be treated. Having positive moral status implies that it is good to benefit an individual and it is ethically problematic to harm them. Negative moral status implies the opposite. It implies that it is good to harm an individual and morally problematic to benefit them. The existence of moral status doesn't rule out this possibility. In other words, the claim that moral status necessarily confers protection assumes there are no moral status decreasing properties; it does not establish that there aren't. Despite that, no individuals have negative moral status.[5] Why not? The capacities just mentioned (for torture, genocide, indifference to suffering) are clearly negative, and those who possess them would be better off if they didn't. But possession of these capacities does not lower their moral status. The best explanation is that our properties do not influence the level of our moral status one way or the other, further supporting the conclusion that there are no moral status enhancing properties.

3.3. Do Capacities Speak Louder Than Actions?

Proponents typically cite one or more *capacities* as the basis for our greater moral status. For example, they cite our capacity for moral agency, that is, our capacity to recognize and act in accord with the principles of morality. This makes sense. We are assuming that moral status enhancing properties, if they exist, make the individuals who possess them morally more important. Presumably, properties that we possess transiently or occasionally can't do that. Central capacities, such as our capacity for moral agency or autonomy, are better candidates. They arguably help to define the kinds of individuals we are. We do not simply have the capacity for moral agency; we are moral agents. This suggests that, if any properties can make us morally more important, it will be properties that help to define who we are.

With this in mind, consider how we define others in moral terms, whether someone is a good person or a bad person. We don't make these determinations based on the capacities the individual possesses. We make them based on what the individual does with their capacities, with their capacities providing a kind of baseline against which we judge their actions. We judge those who do not take advantage of their natural gifts more negatively than those who accomplish little, but did not have the same potential to begin with. We do not, for example, judge negatively people who do not succeed at sports or music if they have no athletic or artistic abilities. This puts in a different light the fact that we have greater capacity for moral behavior compared to animals. We are smarter than rabbits; we are able to recognize the principles of morality and act in accord with them. This provides the baseline against which we judge our actions and the actions of others.

Human beings have done countless wonderful things with our cognitive capacities. These things redound to our credit and support a positive evaluation of human beings. Unfortunately, humans have also done many terrible things. To consider one of the more unfortunate events in human history, World War I has been summarized as follows: "8 million people were destroyed because two persons had been shot.[6] And 8 million dead significantly underestimates the magnitude of the harm. Tens of millions more were injured, assaulted, raped, and abused, and perhaps hundreds of millions suffered trauma, often lasting a lifetime. This represents the harm human beings caused in just four years, from 1914 to 1918. World War I led to World War II, with human beings essentially replaying the horror just twenty years later, this time to even greater effect. Instead of humans killing 8 million people, we killed 50 million.

We noted, in the previous section, that if our positive capacities can increase our level of our moral status, our negative capacities could decrease it. The possibility of reductions in our moral status becomes even more significant when we consider whether our actions might influence our level of moral status. If they can, we would need some way to compare the positive and the negative to determine which is greater. It is certainly not obvious how this might be done and, in the end, how it might turn out. It could turn out that we have lower moral status than animals.

Proponents of degrees of moral status might try to block this possibility by arguing that history raises troubling questions regarding the moral *character* of many individuals and, perhaps, of human beings in general. But our moral status is determined by our capacities, not by our character, hence, not by our actions. To support this response, proponents would need to explain why, even though the capacities we possess can influence our level of moral status, our moral character and the things we have done can't.[7]

Moral character can be understood in terms of the extent to which one has the capacity to identify what one ought to do *and* one does that thing because it is the right thing to do. Our moral character is thus closely connected to a number of the capacities which are typically cited as moral status enhancing, including the capacity to recognize the principles of morality and the capacity to act autonomously. This connection suggests that, if our capacities can influence our moral status, so, too, can our character and, by implication, our actions. Importantly, our character and our actions, like our capacities, help to define who we are. Hence, if the properties which define who we are could influence the level of our moral status, our moral character and actions could as well, raising the concern mentioned in section 3.2 that we might end up with lower moral status than rabbits. The fact that this is not the case points to, and supports the conclusion of, sections 3.1 and 3.2. Our actions and our character do not increase or decrease the level of our moral status because the properties we possess do not influence the level of our moral status.

3.4. The Problem of Relevance

As noted previously, some commentators regard one or more of our superior cognitive capacities as moral status *conferring*: they determine whether an individual has moral status, hence, whether the individual's other properties have moral relevance at all. Whether a rabbit's pain and suffering matter morally depends, according to this view, on whether the rabbit has the cited superior cognitive capacities, for example, whether it is autonomous or a moral agent. The primary problem with this view is that causing individuals to experience pain is morally problematic because it hurts, not because it hurts an individual who is autonomous or has moral agency. This is reflected in our reaction to *House on Fire*. Whether we have reason to rescue the rabbit does not depend on whether it can recognize the principles of morality. The fact that the rabbit will suffer is enough.[8] This points to a significant challenge for the view that our superior cognitive capacities are moral status *enhancing*.

If the moral relevance of a given instance of suffering depends on how bad the suffering is for the individual, whether that individual has superior cognitive capacities appears to be irrelevant. Put in terms of a question: If possession of one or more superior cognitive capacities does not determine whether a given instance of pain matters morally at all, how could it be relevant to determining how important the pain is morally? In both cases, the challenge is the same: the moral relevance of pain depends on the extent to which it leads to suffering in the individual who experiences it.

As we saw in section 2.1, this is the challenge posed by James Rachels. Proponents of degrees of moral status, as we saw there, have a response: they argue that moral status enhancing properties do not directly affect the moral relevance of individuals' experiences or interests. They do not make the pain itself greater or morally worse. Instead, they make the individuals themselves morally more important, which implies that their experiences or interests have greater moral significance. It is morally worse to cause the same amount of suffering in an individual who possesses moral status enhancing properties because they are morally more important. To support the postulation of degrees of moral status, then, proponents need to explain what it means for some individuals to be morally more important than others.

The present question asks how the fact that some individuals possess the cited moral status enhancing properties is relevant to how they ought to be treated, why it makes it worse to harm or fail to respect them. Proponents of degrees of moral status respond that possession of these properties makes the individuals morally more important. This response raises the question of what it means for some individuals to be morally more important. To this question, proponents respond: it means that it is worse to harm them (or fail to respect them). This response reveals that postulation of degrees of moral status is circular: the interests, rights, or claims of individuals who possess the cited superior cognitive capacities have greater moral weight because the individuals are morally more important. And being morally more important involves the individuals' interests, rights, or claims having greater moral weight. This circular reasoning leaves us without an explanation for why some individuals are morally more important.

The absence of an explanation is problematic for proponents of degrees of moral status. It also magnifies the challenge posed by an excess of candidates that we discussed in section 3.1. Humans and animals possess many important cognitive and other capacities (e.g., to fly, reproduce, sacrifice for others). The lack of understanding of what being morally more important involves makes it impossible to determine which of these properties are moral status enhancing. If we don't know what greater moral status involves, we can't figure out what might produce it. From that, it follows that postulation of degrees of moral status offers no way to determine *which* individuals are more important morally, hence, no way to determine whether we are morally more or less important than birds, bats, and immortal jellyfish. These problems with postulations of degrees of moral status support the conclusion that no properties influence individuals' level of moral status. The final step in the argument is to consider why this is the case: Why don't any properties influence our level of moral status? The answer, as we shall see, is that they don't, because they can't.

3.5. The Nature of Thresholds

It is often said that you are pregnant or you aren't. Being married seems similar. People aren't more or less married. Some people are married; others aren't. Other properties come in degrees. Height, for example, is essentially continuous so that two people can vary in height by the smallest degree. Continuous properties pose a challenge when we try to separate individuals into categories. Imagine you are charged with dividing a group of people into married and unmarried. That's pretty easy. Now imagine dividing the same group into tall and not tall. This is more challenging. The identical twin brothers in the group appear to be the same height, even when standing back to back. So you put them in the same group. But the engineer brother, using his fancy laser, shows you that he is 5'10.1" whereas his twin is just 5'10". On that basis, he claims, he is tall and his brother is not. That would be odd. Such a small difference, a tenth of an inch, cannot suddenly make someone tall; it just makes them a tenth of an inch taller.

To consider another example that is relevant to the present discussion, the extent to which individuals experience pain as the result of particular actions, being stuck with a needle as part of medical research, for example, ranges from barely noticeable to significant pain. With respect to moral action guiding properties, this does not seem to pose a serious problem, at least in principle. When we are deciding which of two actions to perform based on the pleasures and pain they might produce, we do not need to divide the potentially affected individuals into groups: those who are sensitive to pain and those who are not sensitive to pain. Instead, we can compare the amount of pain each individual will experience, with the goal of causing less pain rather than more.[9]

Now consider the moral status enhancing properties that are cited as the basis for our greater moral status, such as being autonomous, being a moral agent, and having a narrative identity.[10] All of the plausible candidates come in degrees. The capacity for autonomous action involves the capacity to recognize that one has options for how to act and to choose an option based on one's own preferences. The property of having a narrative identity over time requires a sense of oneself continuing over time and the experiences, plans, and projects being part of one's own life. An individual can be more or less adept at recognizing and understanding the options one has. An individual can have a greater or lesser sense of one's projects and experiences being part of a unified life. The fact that the underlying properties come in degrees reveals that the cited moral status enhancing properties themselves come in degrees. Individuals can be more or less autonomous; they can have a greater or lesser narrative identity. If possession of these capacities increases individuals' moral status, and they come in degrees, it seems to follow that there are essentially countless degrees of moral status depending on the precise extent to which a given individual possesses the cited capacities, just like there are countless different heights.[11]

We have been assuming that whatever level of moral status rabbits possess, they all possess it to the same extent. But this view implies that there could be countless levels of moral status among rabbits. Hence, when conducting medical experiments, the choice is not simply whether it is better to use rabbits or rats; it would be important to identify rabbits with lower levels of moral status.

Similarly, before deciding which human being should get a limited resource, such as a life-saving organ, we would first have to determine the levels of moral status of the potential recipients on the grounds that it is better to give a life-saving organ to someone who matters more morally than someone who matters less.

While that principle is right, tiny differences in the cited cognitive capacities, such as the fact that one person has a slightly greater capacity to see their life as an integrated whole, does not provide a reason to give them a lifesaving organ over someone else. That fact provides strong reason to reject the postulation of countless degrees of moral status. Moreover, our intuitions are not limited to specific actions, such as whether we should save the rabbit or the human being from the *House on Fire*. We also have intuitions regarding the nature of moral status, including the intuition that there are not literally hundreds of millions of levels of moral status among human beings. To see this, consider the individuals in your own family. Even though there are undoubtedly slight differences in the extent to which they possess the cited cognitive capacities, they all have the same level of moral status. These intuitions provide additional reason to reject postulations of countless degrees of moral status.

Finally, endorsement of countless degrees of moral status magnifies the problem of relevance discussed in section 3.4. Proponents who endorse a few levels argue that possession of one or more superior cognitive capacities increases individuals' moral status as a result of making the individuals themselves morally more important. While we rejected that claim, it makes sense. In contrast, the claim that every increase, no matter how small, in a moral status enhancing property makes individuals morally more important does not make sense. It does not make sense to claim that you are more important morally than I am because you possess one superior cognitive capacity to the smallest degree more than I do; for example, you are able to remember one more thing about the past than I can remember.

To address these concerns, while still relying on moral status enhancing properties, some proponents postulate a threshold along the spectrum of the cited capacities. All those whose possession of the cited capacity falls below the threshold have one level of moral status, while all those whose capacity exceeds the threshold have a higher level of moral status (proponents who endorse several levels of moral status endorse several thresholds). Perhaps the most prominent defender of this view is the philosopher John Rawls, who argued that human beings are morally more important than animals because we are moral *persons*, which he defines as individuals who have the capacity to recognize the principles of justice and the capacity to recognize what is good for them personally.[12]

Rawls realized that human beings possess these capacities to varying degrees. For example, recognizing the principles of justice, according to Rawls, involves having a normally effective desire to act according to the principles of justice. The extent to which individuals have this desire comes in degrees. We can imagine, for the sake of discussion, that the points along this range go from 0 to 100. Rocks have no desire to act according to the principles of justice, so they get placed at 0 on the scale. Angels, who presumably have perfect desire in this regard, occupy position 100.[13] Positions 1–19 might be occupied by beings who increasingly have the desire to act according to the principles of justice but who are not able to guide their actions accordingly. Positions 20 and above are occupied by individuals who have the desire and are increasingly able to guide their actions according to the principles of justice. Human beings' capacity in this regard varies so that we might occupy positions 30–75. This suggests that people who occupy position 35 have some moral status and people who occupy position 35.01 have slightly greater moral status, and so on. Rawls recognized what we just learned: this doesn't make sense. Tiny differences like that in our cognitive capacities don't make significant moral differences. He argues instead that whether an individual is a moral person, and has higher moral status, depends on whether they possess the cited capacities to a *sufficient degree.* All those whose capacity exceeds the threshold (e.g., competent adults) are moral persons and have full moral status; all those whose capacities fall below the threshold (e.g., rabbits) are not moral persons.

In the height example, it seemed arbitrary to claim that males who are 5'10" are not tall, but those who are 5'10.1" are tall. We might not be worried about this when it comes to height since it doesn't really matter whether one is categorized as tall or not tall. But the difference in having higher moral status is important. It determines whether one can be subjected to excruciatingly painful experiments, whether one has priority for a life-saving organ. Consider three individuals. One has a desire to act according to the principles of justice that is just below the threshold. The desire of the second is slightly greater to the extent that it is just over the threshold, and the third has the desire significantly more than the first two. The minor increase from the first to the second has dramatic implications, whereas the dramatic increase from the second to the third has no implications as far as moral status is concerned.

This would make sense if one thought that increases beyond the threshold in the capacity that is the basis for our moral status are not important when it comes to morality. If even dramatic increases beyond the threshold had no real significance, we could understand why those increases do not affect one's moral status. But this is not an option for proponents of degrees of moral status. Significant increases beyond the threshold are increases in a capacity that is important morally, the very capacity that provides the basis for higher moral status. Thus, proponents cannot say that increases beyond the threshold are not valuable or important as far as morality is concerned.

This is a significant challenge for all postulations of degrees of moral status which base our greater moral status on properties which come in degrees.[14] John Rawls tried to address it by arguing that the property of being a moral person is a "range" property in the sense that it applies over a range of the underlying capacities.

For example, it might extend from 35 to 100 on the scale. To try to explain how this might work, Rawls appeals to the property of being inside a circle, which applies to all the coordinates within the circle, while Jeremy Waldron mentions being in Scotland. Gretna Green, a city on the border with England, is just as much in Scotland as a city in the middle of the country.[15]

There are many properties like this. Being old enough to drive depends on one's age, which depends on the amount of time that has elapsed between one's birth and the present moment. But there is a clear threshold between those who are too young to drive and those who are permitted to drive—where I live it's seventeen. To take another example, many diseases are defined based on thresholds on the associated symptoms. Depression (or major depressive disorder) is defined as having five or more negative symptoms during a two-week period. There is nothing magical which occurs when one crosses these thresholds. The transition from 16 years and 364 days to 16 years and 365 days does not make one a better driver. An increase in symptoms of depression from 13 days and 23 hours to 2 weeks does not make one significantly worse off.

In these cases, it is widely recognized that the specific point at which we draw the threshold is ultimately arbitrary. We know that three-year-olds do not have the capacity to drive and average thirty-year-olds do. We could set up a system where everyone is tested once a year, or once every six months, to determine when they are able to drive. But that would be burdensome and costly, and there would be endless debate over which decisions were right, so we pick a date, even though we know that the one we choose is no more significant than the ones just before or after it. Similarly, the borders of a circle or a country are typically influenced by a number of factors (presence of a river in the case of a country border). But we also draw these borders based on our own goals and purposes.

When it comes to things like citizenship and laws, how far one is from the center of the country is essentially irrelevant. How much of a claim Scots have to vote is not affected by whether they live in

Gretna Green or Killiecrankie, in the heart of the country. In contrast, as we have seen, the underlying capacities for moral status enhancing properties are morally relevant, and possessing more of them is morally preferable. This is evident in how we describe the cases. Scots who live in the center of the country do not somehow live in Scotland to a greater extent. In contrast, individuals who have a greater understanding of the principles of justice are moral persons to a greater extent. This analysis thus suggests that Rawls's approach does not offer a plausible basis for degrees of moral status.

In the case of the driving age, we recognize that those who are 16 years and 364 days old are no different than those who are 17. But the burden on these individuals is minimal: they just have to wait a day and they will have the right to drive, too. In contrast, the threshold on elevated moral status is critical. It determines whether one can be exposed to painful experiments lasting hours or days to help others and who has priority for life-saving organs. Arbitrarily chosen thresholds cannot justify these practices, revealing that thresholds we set cannot provide a basis for degrees of moral status.

There are other thresholds which are not the result of our choices. The challenge this conclusion leaves us with, then, is to consider how these thresholds work and see whether they might offer a possible basis for degrees of moral status. To consider a prominent example, increases in the temperature of water make it hotter, but it is not until it reaches the threshold that water goes from being increasingly hot to boiling. At sea level, the significant increase from 20°C to 99°C is silent with respect to boiling, whereas the much smaller increase from 99°C to 101°C takes one across the threshold to boiling. Thresholds of this type (the escape velocity of a rocket is another example) have the structure needed for degrees of moral status. Increases below the threshold do not affect one's status, much smaller increases that take one across the threshold change one's status significantly, and further increases beyond the threshold are silent with respect to one's status. Do these thresholds offer a model for the basis of degrees of moral status?

Boiling, which involves liquid water turning into water vapor, occurs when the pressure pushing the water to boil (the internal vapor pressure) exceeds the pressure that is keeping it from boiling (the external atmospheric pressure). The reason why there is a threshold is that there is a point at which the force pushing the water to vaporize exceeds the force that resists its doing so. We can think of these as *competing-factor* thresholds in the sense that they are a result of the resolution of a competition between forces. Because competing-factor thresholds do not trace to choices made by human beings, they offer the potential to avoid being arbitrary.

Moral status enhancing properties, such as one or more superior cognitive capacities, could provide the positive force which increases individuals' moral status.[16] What is needed, then, are properties which have the effect of resisting this force. This reveals that, in order to avoid being arbitrary, postulation of degrees of moral status needs a type of property we have already discussed: moral status decreasing properties. The extent to which an individual possesses one or more moral status decreasing properties (e.g., greed, the capacity for genocide) would provide the force which tends to reduce their moral status, while the extent to which they possess the cited superior cognitive capacities would tend to increase their moral status. The threshold on higher or full moral status would be located at the point at which the positive forces exceed the negative ones.

The first problem with this approach is that the threshold would be located at different places for different people, depending on the extent of their moral status increasing and decreasing properties. For example, individuals who have a greater capacity for greed would need to possess the cited superior cognitive capacities to a greater extent before they gained full moral status. More importantly, postulation of moral status decreasing properties raises the possibility that human beings could end up with lower moral status than many animals and, indeed, we might end up with negative moral status. Endorsement of moral status decreasing properties thus helps postulations of degrees of moral status to avoid being arbitrary. But they do so at the significant cost of raising concern that we might have lower moral status than rabbits and might turn out to have negative moral status.

One might think that proponents could avoid the need for a point threshold and moral status decreasing properties by positing a range threshold instead. In the previous example, rather than claiming that the transition to higher moral status occurs precisely when one's capacities reach 35, proponents might argue that scores of 25–35 represent a gradual transition from lower to higher moral status. Individuals below 25 have the same moral status as rabbits, those between 25 and 35 have an intermediate level, and those above 35 share our greater moral status. Unfortunately, this response does not eliminate the need for a threshold; instead, it replaces one threshold with two: one at 25, which marks the boundary between the lower and intermediate levels of moral status, and another at 35, which marks the transition from the intermediate level to our level. This suggests that postulations of degrees of moral status are not able to address the challenges that arise in attempting to identify a non-arbitrary threshold.

Briefly summarizing the present section, moral status does not come in countless degrees. For example, it is not the case that there are hundreds of millions of different degrees of moral status among human beings. It is not morally better to cause pain and suffering to one human being compared to another because the superior cognitive capacities of the former re lower to an infinitesimal degree. This reveals that there is either one level of moral status or several discrete levels. Discrete levels require non-arbitrary thresholds on one or more moral status enhancing properties. However, there are no non-arbitrary thresholds on the properties which are plausible candidates for being moral status enhancing (e.g., superior cognitive capacities). It follows that there are not discrete levels of moral status, leaving us with the conclusion that there is only one level of moral status. In the end, then, the reason why the properties that we possess do not increase our moral status is because they can't. And they can't because the types of thresholds that are needed for discrete levels of moral status don't exist.

Summary

Many people have the intuition that there are degrees of moral status. Others, such as James Rachels, do not share this intuition. They believe that all sentient individuals have the same level of moral status. To try to resolve this stalemate, the present chapter evaluated what would have to be the case for there to be degrees of moral status. We have seen that degrees of moral status depend on moral status enhancing properties, properties which make the individuals who possess them more important morally than those who have moral status but do not possess the properties (to the same extent). The present evaluation offers compelling reasons to conclude that there are no such properties.

First, views that regard the capacity for moral agency (or some other superior cognitive capacity) as necessary for being a member of the moral community offer a clear reason why that property has special moral significance. In contrast, views which regard these properties as moral status enhancing, rather than moral status conferring, face the challenge of explaining why countless other properties are not moral status enhancing as well (e.g., being selfless, being immortal). The fact that these properties are not moral status enhancing thus suggests that no properties are. Second, human beings possess many negative properties (e.g., the capacity for cruelty). If the properties we possess could influence our level of moral status, possession of these properties could decrease it. The fact that they don't supports the conclusion that the properties we possess do not influence the level of our moral status. Third, in addition to having done many wonderful things, human beings have done many terrible things. Yet this history does not imply that our moral status is lower than the moral status of beings who have not done these things. The fact that our actions do not influence the level of our moral status suggests that our capacities don't either.

Fourth, the problem of relevance suggests that there are no properties which make individuals more important morally in a way that influences a broad range of the ways of treating them. This suggests that no properties are moral status enhancing. Fifth, and finally, the types of thresholds that could provide a non-arbitrary basis for degrees of moral status, what I have called competing-factor thresholds, depend on the existence of moral status decreasing properties, which even proponents of degrees of moral status reject.

This analysis reveals that there are no moral status enhancing properties. And since degrees of moral status depend on the existence of such properties, it follows that there are no degrees of moral status. You either have moral status or you don't. Of course, not everyone will automatically agree. When scientists started noticing things that conflicted with the view that we inhabit the center of the physical universe, they didn't immediately reject that view. Instead, they described ever more complex theories of how the planets move to explain what they saw. This response was supported by several factors.

First, people were used to the theory that the Earth is located at the center of the universe and they were comfortable with it. Second, it made sense of a good deal of ordinary experience. Third, it has the reassuring consequence of placing us at the center of the universe. What finally brought the process of proposing increasingly complex theories to a halt was the description of an alternative view—the sun is at the center of our universe—which better accounted for what scientists saw. At the same time, this view did not instantly resolve all cosmological questions and puzzles. To the contrary, many questions and concerns were answered only after scientists took this view seriously and sought to understand it.

The present challenges facing moral status enhancing properties bring us to a similar place with respect to the moral universe. Despite these challenges, some will continue to believe that we are more important morally than animals. We are used to this view. It has a long history. It has the reassuring consequence of placing us at the center of the moral universe. And this view is consistent with strong intuitions that we have in response to cases like *Experimentation* and *House on Fire*. But the present chapter reveals that that is not the world we live in. This conclusion, like the conclusion that the Earth is not at the center of the universe, does not instantly reveal a full and fully comprehensible account of how we should act. Instead, this conclusion reveals that we live in a world without degrees of moral status and leaves us with the challenge of investigating and understanding it. How do we make sense of a world without degrees of moral status, of a moral universe without us at its center?

To begin this assessment, we will return to the discussions from Chapter 1, where we considered the fact that, intuitively, it seems worse to kill human beings, to cause them to suffer, and to fail to respect them compared to animals. These intuitions offered important support for degrees of moral status: animals matter morally, but we matter more than they do. They thereby provide reason to be skeptical of the present conclusion that there are no degrees of moral status. One way to assess this conclusion, then, is to assess whether we can make sense of these intuitions in ways that do not appeal to degrees of moral status. A finding that we can would further support the present conclusion. A finding that we can't would undermine it. That is the challenge for Chapters 4, 5, and 6.

4
Is It Really Worse to Kill Humans?

Introduction

Chapter 3 concluded that there are not any properties which increase the moral status of those who possess them. Given that degrees of moral status require moral status enhancing properties, it follows that there are no degrees of moral status among those who have it. To explore this conclusion, the remainder of the book considers life without degrees of moral status. What are the implications of the conclusion that there are no degrees of moral status? To begin this assessment, the present and next two chapters return to the three intuitions with which we began: it is morally worse to kill, to cause to suffer, and to fail to respect human beings compared to animals. The challenge for these three chapters is to see whether we can make sense of these intuitions in a world without degrees of moral status.

A great deal has been written on the topics we will cover in these three chapters.[1] However, the present goal is not to summarize and assess what has been said before us. The goal is to see whether we can make sense of these intuitions without degrees of moral status. For that purpose, I will discuss my own thoughts and analysis, along with the relevant thoughts and analyses of others, without offering anything close to comprehensive coverage of their views.

Life Without Degrees of Moral Status. David S. Wendler, Oxford University Press.
DOI: 10.1093/oso/9780197675328.003.0005

4.1. Killing, Dying, and Being Dead

Given a choice between killing a human being and killing a rabbit, the morally right option seems clear: it is significantly worse to kill a human being than it is to kill a rabbit. The question for the present chapter is whether we can explain this difference without degrees of moral status. If we can't, it follows that either there are degrees of moral status, or there is not a significant moral difference between killing human beings and killing rabbits. And, given that it is very problematic to kill human beings, the latter possibility would suggest that it is very problematic to kill rabbits as well, even painlessly.

Killing an individual involves two things. There is the death of the individual who is killed, and there is the something or the someone, the killer, who causes or brings about their death. To evaluate whether it is worse to kill a human being than it is to kill a rabbit, we need to evaluate both aspects. Is it worse for a human being to die than it is for a rabbit to die? Is it worse to cause or bring about the death of a human being than it is to cause or bring about the death of a rabbit? I will argue that death and killing *are* worse when it comes to human beings compared to rabbits. But this is not because we have greater moral status. Understanding why not provides valuable insights into the nature of death and killing, insights which can be obscured if we assume these differences trace to the fact that we are morally more important than animals.

4.2. Why Is It Bad to Die?

Dying involves the transition from being alive to not being alive, or not existing. One might think that dying is bad, then, because it involves moving from a good state, being alive, to a worse state, being dead. To assess this possibility, consider someone who died two hundred years ago. Do you think they are now in a much worse situation compared to when they were alive? That doesn't seem right.[2] Being dead is not terrible for the individual who is dead for the simple reason that death does not simply eliminate the life the individual was leading; it eliminates the individual who was leading that life. Death cannot be bad for the person, or so it seems, because a person who is dead no longer exists, in which case nothing can be bad for them. This line of thinking led the ancient Greek philosopher Epicurus to argue that death is not bad for us because, when death arrives, we are no longer around to experience it.[3] Experiencing pain is bad for us because it hurts; it leads to suffering. But nonexistence isn't like that. It involves the absence or the elimination of all experience; hence, it does not involve any experiences at all. Nonetheless, it still seems bad for us to die. Even people who do not believe in a ghastly afterlife, or any afterlife at all, fear death. Why?

Earlier we discussed the challenge of trying to evaluate our position in the universe while standing on the Earth. In that case, we tried to keep in mind that our perspective might be influencing what we think. Even after we realize the Earth is moving around the sun, it still seems like the Earth isn't moving. A similar possibility arises here. When we think about whether, and to what extent, it is bad for us to no longer exist, to be dead, we are necessarily considering that possibility from the point of view of someone who is alive. In at least two ways, that perspective can influence our views on death.

First, our evaluation of the badness of being dead is not influenced only by how bad it is for the individual who is no more. We are also influenced by how bad it is to make the transition from existing to not existing. In particular, our evaluation of the badness of death is affected by the fact that it involves leaving behind everything we know and everything we care about. To appreciate how this can influence what we think about death, consider moving to a new home. Moving is one of the more stressful experiences in the lives of many people, even those who recognize that they are moving to a better place. A primary reason for this is that we tend to focus on what we are leaving rather than where we are going. This makes sense; we often know what we are leaving behind, whereas we don't know, or aren't sure, what life will be like where we are going.

Similar things happen to many people at the beginning of extended trips. I love traveling, but I have the same reaction before every trip. I always feel a little depressed; I think the trip will be bad, that it will be boring and uninteresting and scary. And this happens even when I am going to places I have been before, places I know I will enjoy. The experience of making the transition, the need to say goodbye and leave, affects my evaluation of the destination. Importantly, and unfortunately, my realizing this fact has not eliminated the effect. I continue to think the places I am going will be boring even when I know they won't be. A similar challenge arises when we evaluate the badness of death. We can't imagine being dead for the simple reason that imagining at all depends on our being alive. Hence, our evaluation of the badness of death tends to focus on the transition rather than the end state; we evaluate dying rather than being dead.

Our perspective can influence our views on the badness of death in a second way as well. In the introduction to Chapter 1, we considered whether it is better to be a happy pig or an unhappy person. Although being an unhappy person seems better, that impression comes from the perspective of being a human being. As a result, we aren't really imagining what it is like to be a pig; we are imagining what it would be like for us to be pigs, to be stuck in a pig's body, living a pig's life. Trying to imagine how bad death is for the person who dies poses a similar challenge. I try to imagine being dead while I am alive and kicking and actively thinking. I thus end up imagining myself in a dead person's body, which seems like being trapped in a coffin. Or I imagine being separated from, and observing all the things I care about, and all the things I am attached to. They keep on going, while I am blocked from them by a thick plate of soundproof glass. I see all the good experiences I would have had in the future, all the beautiful sunsets, hot showers, and cold beers. And I miss them.

Of course, those things will go on without me. But I won't be hovering in the sky, watching it all happen. I simply won't be. All of this suggests that it can be difficult or impossible for us to imagine what it is like for a dead person to be dead. Being dead isn't like being in pain. And it's not even like staying home sick and watching out the window as the world passes you by. Death is not like anything for the person who is dead because the arrival of death coincides with their nonexistence. So how can that be bad for them?

The answer many philosophers give is that death is bad for the person who dies, even though nonexistence isn't, because death deprives us of all the goods we would have enjoyed if we had kept on living. It deprives us of all the times we would have had dinner with friends, listened to music, taught our grandchildren how to bake, and watched beautiful sunsets. This is the *deprivation* account of the badness of death.[4] Of course, death also deprives rabbits of everything they had and everything they otherwise would have had. There are, then, at least two ways in which the death of human beings and the death of rabbits are similar: they both lead to the nonexistence of an individual who had moral status, and they deprive that individual of all the good experiences they otherwise would have had.

One might thus try to explain why death is so much worse for human beings by pointing out that humans live on average about seventy-two years (although it varies significantly between richer and poorer countries), whereas domesticated rabbits live on average about ten years. The total amount of life, and the number of valuable experiences lost is, therefore, much greater when the average human being dies compared to the average rabbit. While this makes sense when it comes to rabbits, it does not work for giant tortoises, which live two hundred years on average. To the extent that simply being alive is a significant benefit, the death of the average giant tortoise would be worse than the death of the average human being. In addition, it might turn out that giant tortoises relish every day of life and experience great pleasure floating in the ocean. That would suggest, on the deprivation theory, that the death of giant tortoises is much worse than the death of average human beings. Or, even more extreme, consider the character I introduced in section 3.1, *Turritopsis dohrnii*, more commonly known as the immortal jellyfish, which is very small and lives in the Mediterranean and around the islands of Japan. More importantly, it appears to be immortal.

To understand this, and its implications for the present discussion, it is important to distinguish between being *immortal* versus being *indestructible*. An indestructible being is one who will not die, no matter what happens to it. Drop a house on its head, run over it with a tank, heat it to 10,000 degrees, freeze it to −150 degrees, and it keeps on living. Immortality is different. An immortal being is an individual whose lifespan is *potentially* infinite, provided seriously bad things do not happen to it along the way. In other words, an immortal being is one who will not die for what we might think of as internal reasons. Their bodies and minds do not age and decay to the point where they eventually cease to be. Instead, if things in the environment cooperate (no houses fall on their heads, no tanks come their way), immortal beings can live forever.

While *Turritopsis dohrnii* is not indestructible, it appears to be immortal. If it is left in peace, it will not die. This suggests that killing an immortal jellyfish potentially deprives it of an infinite future,[5] whereas killing the average adult human being deprives them of thirty to forty years of life. It seems unlikely that immortal jellyfish are sentient. And given that scientists have not observed them forever, we cannot be certain that they actually are immortal. Still, whether it is worse for a human being or an immortal jellyfish to die does not depend on the answers to these questions. Imagine scientists discover that the immortal jellyfish is immortal and that it has a very basic psychology, which involves pleasant experiences as a result of living in the ocean. Even in that case, the death of an immortal jellyfish would not be significantly worse than the death of a human being.

One might think that the way to explain this is by appeal to degrees of moral status. Immortal jellyfish might live a lot longer than we do, and have potentially an infinite number of pleasant experiences, but we are morally more important than they are. The amount of goods we lose by dying, factored by our greater moral status, makes our deaths worse. However, even if we are 100 or even 100,000 times more important morally than immortal jellyfish, the fact that they have a potentially infinite lifespan suggests that, on a deprivation account, their deaths will still be worse. No matter how much greater our moral status, as long as it is a finite difference, multiplying the number of positive experiences death deprives us of by that amount will always be less than the total positive experiences of which immortal and sentient jellyfish are deprived.[6]

This conclusion is important. We have been considering the possibility that we need to appeal to degrees of moral status to explain why the death of human beings is worse than the death of rabbits and other animals. However, differences in degrees of moral status cannot explain the fact that the death of the average human being is much worse than the death of immortal and happy jellyfish. Moreover, the reason why our deaths are worse than the deaths of rabbits, and many other animals, is likely to be the same as the reason why our deaths are worse than those of immortal jellyfish. If that is right, appeal to degrees of moral status doesn't explain why our deaths are worse than the deaths of rabbits and other animals. To develop a better explanation, we need to distinguish between two different types of future goods that death deprives us of.

Very generally, our lives involve two things: what *happens to us* during our lives and what *we do with* our lives.[7] How well our lives go for us, for our own sake, depends on both of them. It depends on our experiential interests, which involve whether the things that happen to us are positive or aversive. And it depends on what I will call our "contribution" interests, which involve the extent to which we contribute to valuable projects and relationships (and avoid contributing to detrimental ones). Many activities promote *both* our experiential interests and our contribution interests.[8] Making pottery is a good example. It can be relaxing and therapeutic while we are doing it, thereby leading to positive experiences. Making pottery also involves engaging with a craft that is thousands of years old and honing one's skills. And, at least sometimes, it leads to a nice piece of pottery. To these extents, making pottery promotes our contribution interests.

Or consider being a parent. For many people, raising children reduces their happiness overall; they worry more, they get into more fights, and they have less time and money for themselves. Does it follow that it was a bad idea for them, for their own sakes, to have children? Only if you think that how well one's life goes depends on only the nature of their experiences. For many parents, the primary personal benefit of having children is contributing to the life and flourishing of another person and, through them, to the betterment of others and the world. In this way, raising children can promote parents' contribution interests even when it sets back their experiential interests.

When we are living, we want to have good experiences and we also want to contribute to valuable projects, activities, and relationships. That's what makes a better life for us. As a result, while we are alive, we emphasize the importance of having and continuing to have good experiences. From this perspective, death seems terrible because it robs us of the valuable experiences we would have had in the future. That supports the deprivation account of the badness of death. But this raises the question of why things shift when we consider the death of the immortal jellyfish. Why don't we worry so much about the infinite future pleasures that are lost with the death of a (happy) immortal jellyfish? The answer, I suggest, is that it shifts our perspective. When we think about our own deaths, we inevitably think about them from inside the lives we are living, and the lives that will be lost. But when we evaluate the possibility of the death of an immortal jellyfish, we evaluate it from outside the life in question. That changes things in a critical way.

Evaluating lives from the outside reveals that positive experiences are *contingent* goods: they are good for us while and to the extent we are alive. As long as we are alive, it is better to have nice meals and watch beautiful sunsets. Losing those things by dying seems very bad for the simple reason that we are necessarily alive when we consider the possibility. The question of why it is bad to die is similar to the question of why it is bad to leave your friends and go on a trip. It is bad because you are leaving important goods behind. We recognize that while we are dying. That is the transition. But death is different. It is not just that we won't be with our friends anymore. We won't be anymore. At that point, the question of whether the individual has positive or negative experiences no longer applies because the individual no longer exists.

One way to understand this is to recognize that our experiences come into the world with our existence, and they go out of it with our deaths. They are thus good for us to the extent that we are alive. Put differently, good experiences are good for the individual who has them to the extent that they are alive. Their value depends on the existence of the individual, and their value dies with the individual's death. Once they die, there is not anyone or anything for whom those experiences are good or valuable.

Making contributions to valuable projects and relationships is different. The things we contribute to and achieve can have effects beyond us, and beyond our lives. Our contributions and achievements can make the lives of others better, and make the world a better place. Our contributions thus provide a way to transcend our subjective experiences, our lives, and our deaths. A life that is limited to pleasant experiences ends with us; a life that includes contributions and achievements lives on in those achievements, and in those who are influenced by them.[9] Unlike contingent goods, then, the more achievements a life includes the better.

We are now in a position to explain why death is much worse for the average human being, who loses forty years of life, than for the immortal jellyfish that potentially loses forever, even if the immortal jellyfish is sentient and experiences many pleasures. Death is bad to the extent that the loss of the individual reduces or eliminates their projects, activities, and relationships.[10] And it is bad for the individual to the extent that it sets back their interests in making contributions and having achievements.[11] The death of an average human being interferes with more achievements than the death of a sentient immortal jellyfish. This is true even if the sentient, immortal jellyfish has basic psychological connections across its pleasurable moments—perhaps it remembers the past ones, looks forward to future ones, and compares them. As long as it is limited to pleasant experiences, the goods in its life are contingent goods, goods that matter morally to the extent, and only to the extent, that it is alive.

The death of the average human being thus eliminates more projects and potential achievements.[12] Moreover, we have a much deeper relationship to our projects. We don't simply carry them out. We plan them, anticipate them, prepare for them, and review them. We become invested in them and embrace them as important parts of our lives. They thus play a significant role in the value and meaning of our lives. It is not so much that humans put more effort into their projects; it's that we put more of ourselves into them. In the final chapter, we will consider whether some animals might have similar relationships to their projects and activities.

This explanation for why the death of human beings is worse for them than the death of animals is for the animals appeals to the general principle that it is worse to lose more important and valuable contributions compared to less important and less valuable contributions. And it is worse to cut off the projects of individuals who are more invested in them compared to individuals who are less invested in them. This provides grounds for saying that the death of human beings is much worse for them than the death of rabbits is bad for the rabbits that die. This also shows that more of value is lost with the death of human beings than rabbits. In this way, the fact that the death of human beings is worse than the death of rabbits does not provide reason to believe in degrees of moral status.

This brings us to the second aspect of killing: causing or bringing about the death of the individual. Does the intuition that it is worse to kill human beings point to a difference in degrees of moral status? And if death is not bad for rabbits, for their own sake (it might be terrible for the rabbit's offspring), does it follow that killing rabbits painlessly is not bad at all?

4.3. Why Is It Wrong to Kill?

The previous section concluded that the painless death of a rabbit is not a very bad thing for the rabbit itself. Perhaps more surprisingly, a painless death is not nearly as terrible for us as we imagine. Death deprives individuals of all they had and all they would have had. But almost all of the valuable things in the lives of rabbits are contingent goods; they are good for the rabbit to the extent the rabbit is alive. And many of the things death deprives us of, in particular all our positive experiences, are similarly good for us to the extent we are alive.

Given that painless death is not all that bad for the individual who dies, one might be tempted to conclude that painlessly killing others is not that bad either (reminder: we are focusing on the extent to which killing is bad for the individual killed; there are many ways in which killing an individual can be bad for those who care for or depend on them). This makes sense to the extent we assume that the wrongness of killing an individual depends only on how bad it is for the individual to be dead. Put differently, if nonexistence is not bad for the individual who doesn't exist, any explanation for why it is wrong to kill them will have to cite other factors. But what other factors might there be beyond the individual being dead?

To answer this question, it is helpful to think about killing an individual in terms of destroying them. The badness of destroying something does not depend solely on how bad it is for the thing to be destroyed for its own sake. Consider destroying a great work of art or a great work of nature. To take a specific example, on May 21, 1972, a (mentally disturbed) geologist walked into the Sistine chapel and attacked the Pieta, a world-famous sculpture and the only piece Michelangelo ever signed. The Pieta was not made worse off, for its own sake, as a result of the attack. Still, it was a seriously problematic thing to do. Why? Because the Pieta possesses significant aesthetic and historical value. Not only did attacking it fail to show respect for its value, it reduced its aesthetic value.

While human beings and rabbits do not have great aesthetic value, they have value in other ways. In particular, human beings and rabbits are complex organisms, and the destruction of them involves the destruction of a complex organism.[13] This provides a reason why it can be problematic to kill human beings and rabbits which does not trace to the moral status of the individuals. Moreover, when individuals have projects and plans, when they have preferences for how their futures go, they merit a kind of respect which involves permitting them to carry out and realize those plans and projects. Killing the individual obviously fails to respect them in this way. Killing an individual involves deciding for them how the rest of their life goes. This reveals that, although they are closely connected, the badness of death and the badness of killing involve different things.

Death involves the nonexistence of an individual, one who no longer has experiences or a life. Killing, in contrast, does not involve individuals who no longer exist. Killing is not something you can do to the dead. You can only kill the living, thereby deciding for them how their lives will go, and how they end. Killing, in this sense, involves perhaps the ultimate exercise of domination of one individual over another. This explains why killing someone who is old is about as bad as killing someone who is young; why killing the morose is about as bad as killing the contented:

> It is, according to the equal wrongness thesis, no less seriously wrong to kill an old person than to kill a young person, or to kill a person of melancholy disposition than to kill a person with a happy temperament. The extent to which it is wrong to kill a person is, in short, unaffected by the degree of loss or harm he would suffer by being killed.[14]

This raises a puzzle: how can it be seriously wrong to do things to individuals that do not make them seriously worse off?

To see how, imagine a sensitive thief who wants to ply their trade without seriously harming anyone. They thus devise a plan to electronically steal 500 dollars from the bank accounts of each of the 2,600 billionaires in the world. Carrying out this plan would result in the thief gaining 1.3 million dollars, while each billionaire loses 0.0000001% of their worth. Most people who suffer such a loss wouldn't even notice it, and there is essentially nothing the billionaires could do before the theft that they can't do after it. This suggests that the theft does not make the billionaire victims worse off; it does not harm them. Hence, if the ethics of how we treat others depended solely on the impact on them, whether it leaves them worse off, it would follow that the theft of this money is not morally problematic. Yet stealing even a small amount of money from someone else's bank account is, in standard cases, seriously wrong. One piece of evidence for this is that someone who is guilty of such a theft should be punished. Yet if the crime of stealing 500 dollars from billionaires has essentially no negative impact on them, why should the thief be punished?

The straightforward answer is that taking things which rightfully belong to others without their permission violates their right to decide how their lives go, in this case, what they do with their money. Stealing 500 dollars from billionaires doesn't simply diminish their worth by 0.0000001%; it violates an important right they have over their money by deciding what happens to their bank account. It wrongs them even when it doesn't harm them. While this answers the present question, it raises a further puzzle. Why would stealing 500 dollars from billionaires violate their rights, and thereby wrong them, when it does not have a significant impact on them? Why do individuals have rights against being treated in ways that do not make them noticeably worse off?

The answer traces to the importance of our having a sphere of influence over our lives that is protected from the intrusions of others. This involves at least two things. It involves others not determining how our lives go for us, and it involves our being in a position to determine how our lives go for ourselves. This is sometimes described in terms of the importance of respecting individuals' autonomy. Human beings have the capacity to shape the course of their own lives, and we have a significant interest in being permitted to do so, absent the intrusions of others.

To consider a second example, one which brings us to the ethics of killing, imagine a seventy-year-old man with a terminal illness. For four months, he works on completing the projects he was pursuing and spends time with and says goodbye to those he loves. He now has one week to live. He is somewhat anxious and depressed, but not in pain, and he is able to care for himself. Now imagine that a stranger sneaks into the man's room and kills him painlessly, while the man is asleep. Analyzed in terms of what the man has lost, the extent to which his life has been shortened, and the experiences and goods he thereby loses, what the stranger does isn't all that bad. Comparing the two lives, the life that lasted seventy years versus the life that lasted seventy years plus one week, suggests there are essentially no differences in the net goods for the victim. Does this suggest that what the stranger did was not wrong? No. The wrongness of killing does not depend simply on the extent to which it deprives the victim of goods they otherwise would have experienced. It depends on who does the depriving. The stranger decided when and how the man would die. Not only did the stranger not have a right to do that, but the man himself had a strong claim to direct the course of his life.

One way to see this, and to gain insight into the extent to which the ethics of death and the ethics of killing involve different things, is to change the story. Imagine that, rather than being killed by a stranger, the man with one week to live is killed by a tree that falls on him, killing him instantly, while he was taking a walk in his favorite park. In this case, his death does not seem so terrible, and we might even imagine some people considering it to be a small blessing. But notice that the impact on the man and his life is the same in the two cases. The difference is that other people can violate our right to direct the course of our lives; trees can't do that, even when they kill us. The explanation, then, for why it is so bad to kill human beings, including human beings who have little time and no projects left to complete, is that they have a right to determine how their lives go and how they end. This reveals that the explanation for why it is so bad to kill human beings does not depend on and thereby does not support degrees of moral status (in section 6.2, we will consider whether animals have a right not to be killed).

Summary

We began this chapter with the fact that killing human beings is morally much worse than killing rabbits, asking whether this difference points to a difference in degrees of moral status. We noted the possibility that our perspective, being alive, might influence our response. While we cannot address this potential confounder by thinking about the badness of death while we are in some other state, we can consider death and killing from two different perspectives. The first is from within the life. How do we think about our lives, from the inside, while we are living them? From that perspective, death seems terrible, horrible, because it deprives us of everyone and everything we have. At the same time, the fact that we are assessing the badness of death while we are alive obscures the fact that many of the things we lose are things which are good for us to the extent that we are alive.

Most of what rabbits have and lose when they die are contingent goods in this sense, suggesting that a painless death is not a very bad thing for rabbits. And not that much is lost when a rabbit dies. Given that a large part of why we think death is so bad for us concerns the loss of contingent goods, loss of our positive experiences, our deaths are not nearly as bad for us as we think. The fact that we recognize this, at least to a certain extent, is illustrated by a common reaction to the death of someone who has lived a full and productive life: we often feel sorry, not for them (even though they thereby missed out on future positive experiences), but for us, for the ones who are left to live on without them. This conclusion raised the question of why killing human beings is seriously wrong, even though death is not so bad for most of us.

How bad death is for us depends on the extent to which it sets back our contribution interests. But death does not always set back the individual's contribution interests. This possibility was illustrated by the man with terminal cancer who has one week to live. Why, then, is it so wrong for the stranger to kill him? Killing someone involves deciding the end of their lives for them. It involves thereby subjecting them to our decisions, to our wills, in a fundamental way. The fact that people have a right to direct the course of their own lives reveals that we can explain why it is so wrong to kill human beings without having to believe in degrees of moral status. In other words, even in a world without degrees of moral status, it is seriously wrong, in standard cases, to kill other human beings.

This leaves us with the question of whether it is problematic to kill rabbits. We will return to this question in section 6.2. But before we do, we need to consider the second type of fundamental harm: Is it really worse for aversive experiences, such as pain and suffering, to occur in human beings than in animals?

5
Is Suffering Really Worse in Humans?

Introduction

Chapter 4 concluded that it is worse to kill human beings than it is to kill animals. But this difference does not provide evidence that we are morally more important than animals. Instead, the death of human beings harms them more, by setting back their contribution interests, than the death of animals harms them (in Chapter 8 we will consider whether animals have contribution interests). In addition, killing human beings, at least in standard cases, violates their right to decide the course of their own lives (we postponed until Chapter 6 whether animals have such a right).

The conclusion that the ethics of killing does not provide support for degrees of moral status brings us to cases where the harms between human beings and animals, what they have at stake, are similar. *Experimentation* was designed for this purpose. As a reminder: a researcher identifies a potential new treatment for cancer. The initial study will cause significant pain to five subjects for four hours, and there is no chance that being in the study will benefit them. The amount of pain, and the scientific value of the study, will be the same whether the researcher enrolls human beings or rabbits. What should the researcher do?

Answering this question is complicated by the fact that the properties proponents cite as the basis for our higher moral status frequently influence our experiences, including our experience of pain.[1] For example, adult humans can consent to being enrolled in

Life Without Degrees of Moral Status. David S. Wendler, Oxford University Press.
DOI: 10.1093/oso/9780197675328.003.0006

the experiment. Not only is that morally relevant in itself, but voluntarily accepted pain tends to produce less suffering than imposed pain. On the other hand, obtaining consent requires explaining the experiment ahead of time, which alerts human beings to the coming pain. They might thus experience anticipatory anxiety, increasing the extent to which they suffer compared to rabbits. Moreover, individuals, whether rabbit or human, do not return to their psychological baselines the instant an experience of significant suffering ends. Instead, the suffering tends to linger in the form of negative psychological consequences, stress, anxiety, and, in the extreme, symptoms of posttraumatic stress disorder (PTSD). These effects are likely to be different between humans and rabbits.

For the purposes of assessing whether there are degrees of moral status, we need to set these possible differences to the side and focus on just the experience of intense pain for four hours in five individuals. Is it worse, morally, for those experiences, that level of pain and suffering, to occur in human beings rather than rabbits? Many people think it is morally worse. And, since we are stipulating that the pain experiences are equally bad in the two cases, this intuition points to the possibility that there is more to the ethics of aversive experiences than just the badness of the experience itself; there is the individual who has the experiences, and how important they are morally. These intuitions thus seem to support the claim that we are morally more important than animals.

To assess this conclusion, the present chapter discusses the nature of our perspective and its influence on our intuitions concerning pain and suffering, and other aversive experiences. We will consider several reasons why humans tend to downplay the aversive experiences of individuals who are physically different from us. These considerations raise the possibility that, when we consider examples like *Experimentation*, we might be assuming, without realizing it, that rabbits are less sensitive to pain and suffering; hence, their experiences in the experiment won't be as bad as ours. Because we cannot exclude that possibility, our intuitions regarding pain and suffering (and other aversive experiences) are not reliable indicators of whether it is worse to harm human beings than it is to harm animals to a similar extent. The present chapter thus concludes that these intuitions do not undermine the conclusion that there are no degrees of moral status.

5.1. (Not) Feeling the Pain of Others

In *Experimentation*, I stipulated that the pain would be the same whether the researchers enroll rabbits or humans. That stipulation was useful for assessing whether the moral significance of the same level of pain and suffering depends on the individuals in which it occurs. But, of course, in real life we can't know whether they are the same, or even similar. It is impossible for us to experience or know what it is like to be an animal. It is impossible for us to experience their pain, and it is impossible for us to know what it is like for them to experience it (for discussion, see Box 1.1: What Is It Like to Be a Bat?). For all we know, the rabbits' experience of pain in the experiment might be dramatically worse than the pain human subjects would experience. Perhaps rabbits are extremely sensitive to pain.

Our first dog, Katy, was a 170-pound English Mastiff who tore the ligament in her knee and, years later, developed bone cancer. As part of her treatments, she received many injections and other invasive procedures, in response to which she frequently didn't move or make a sound. The veterinarian was amazed and thought that painful procedures just didn't bother her like they did other dogs. He might have been right, but we worried that she might be experiencing the pain, but not expressing it, and insisted that he give her pain medication, just in case.

You might think that this is how we would typically respond to the experiences of animals. We would assume the pain and suffering they experience in medical experiments might be severe, thus giving us a reason to do our experiments in humans instead. In fact, we don't do that. Instead of assuming that negative experiences we can't access might be especially aversive, we tend to minimize them. This phenomenon is evident in our reactions to war and natural disasters. We take the suffering that they cause more seriously when we have direct access to it, or we hear from those who do. The problem, in the present case, is that we cannot talk to animals that are used in medical experiments. By the way they behave, we can see that they try to avoid painful stimuli, and we can infer that they do not enjoy it. We can also measure their physiological reactions to pain, the extent to which their heart rate increases, for example. But we cannot be sure what subjective experiences are associated with these findings.

In response to this uncertainty, we tend to assume that the pain animals experience is not that bad. This phenomenon provides a possible explanation for our intuitions in response to cases like *Experimentation*. Even when we tell ourselves to imagine that the pain experience in the rabbits is essentially as bad as the pain experience in the human beings, our intuitions might be influenced by the assumption that it isn't. That is, the fact that we know significantly less about the aversive experiences of rabbits might lead us to assume that the experiences are not so bad for them; at least they are not as bad as they would be for us. And that assumption might be influencing our intuition that it is better to conduct the experiment in rabbits. This does not show that it is just as bad for the pain to occur in rabbits. Instead, it provides reason to be skeptical about the reliability of the intuition that the same pain is morally worse when it occurs in human beings.

5.2. Downplaying the Suffering of Others

The previous section considered the human tendency to minimize suffering when we don't have reliable access to it. A related tendency is to assume that individuals who are different from us in superficial ways are different from us in more fundamental ways. We tend to assume that human beings who have different skin color, different hair texture, and different eyes and noses experience less suffering in response to similar experiences.[2] We think that needle injections don't hurt them as much as they hurt us. If such superficial differences can lead to our assuming that other humans do not suffer as much as we do, more substantive differences between us and other species might have even greater effects.[3] If the fact that another human being has slightly darker skin can lead us to assume that they do not suffer as much as we do, it would not be surprising if we assume that individuals who have floppy ears, are covered with dark fur, and hop suffer less than we do from similar experiences, and might suffer substantially less.

This provides more evidence that our intuitions regarding the pain-inducing experiment might not provide evidence for our having greater moral status. Instead, the fact that rabbits are so different might lead us to assume that they do not suffer like we do. Hence, while doing the experiment in us would lead to significant suffering, we might be assuming, without realizing it, that the experiment will result in less suffering in rabbits. That possibility offers further reason to be skeptical of the intuition that the same pain is morally worse when it occurs in human beings.

5.3. Assuming Animals Are Less Important

When it comes to other human beings, psychology experiments find that the extent to which we believe they have moral status, and should be protected from harm, is influenced by the extent to which we perceive them as possessing certain traits:[4] "Viewing others as lacking important elements of humanness has implications for whether they will be seen as deserving protection from harm."[5] In one experiment, individuals who were perceived as having greater interpersonal warmth and emotional responsiveness were perceived as having greater moral status and deserving greater protection.[6] Other experiments find that this effect influences our views of animals as well. For example, participants who were shown images of dogs consistently preferred the images that had been modified to have more human-like features.[7]

These experiments suggest that the tendency to regard human beings as having greater moral status than animals might be essentially built into the way we perceive others. Specifically, the more an individual possesses human traits, the higher we think their moral status is. This provides additional reason to think that our intuitions regarding *Experimentation* might not provide evidence that we are morally more important than rabbits. Instead, these intuitions might be a result of our assumption that human beings have greater moral status.

5.4. (Not) Ignoring the Consequences

Whenever we consider specific examples, and our intuitive responses to them, *Experimentation* and *House on Fire* included, we need to be careful to identify which features of the examples are influencing our reactions and intuitions. This is true in all cases, but especially challenging in the present ones. The goal of *Experimentation* is to keep everything the same between the two versions, except for whether the subjects are humans or rabbits, and consider whether that difference alone makes a moral difference.

So we stipulate that we are considering only the moral significance of the pain and suffering that the experiment triggers and ignoring any possible downstream consequences. However, we can't be certain that our intuitions are playing along. When we picture the experiment being conducted in human beings, we might inevitably worry about the impact it will have on the subjects' education and careers. When we consider the rabbit version of the experiment, we might assume those consequences do not apply. This provides further reason to be skeptical of the reliability of our intuition that the same unit of pain is morally worse when it occurs in human beings.

Summary

The present chapter considered several aspects of our perspective which influence our assessment of the aversive experiences of others. In particular, we have seen that we tend to assume, often without realizing it, that: pain and suffering which we do not have access to isn't that bad; individuals who are physically different from us do not suffer as much as we do; and individuals who are physically different from us have lower moral status than we do. These considerations reveal that the intuition that it is better to conduct the experiment in rabbits is not a reliable indicator of the truth; hence, it does not provide reason to question the present conclusion that there are no degrees of moral status. This is not to say that it is, in fact, no better to conduct the experiment in rabbits (we will take up that question in section 8.5). The present conclusion is more modest than that: we cannot tell whether our intuition that it is better to conduct the experiment in rabbits is influenced by biased factors. Hence, this intuition does not provide reason to question the present conclusion that there are no degrees of moral status.

Of course, we need to take into account the potential for bias in other cases as well. For example, our biases might be influencing our intuitions that it is worse to kill human beings than rabbits, or that it is more important to respect human beings. Granting that possibility, the influence of our perspective is more fundamental and more difficult to estimate or avoid when it comes to assessing the aversive experiences of animals. Our perspective is essentially designed to emphasize the suffering of human beings and to downplay the suffering of individuals we cannot know as well and individuals who are physically different from us. We are not to blame for this, and it might not be an accident. The tendency to think that human suffering is especially bad, and the suffering of animals is not, might have had important evolutionary benefits for humans. A species that does not take the suffering of its members seriously would be much less likely to survive. And downplaying the suffering of animals might have increased our ancestors' willingness to exploit animals for their own benefit, enhancing their survival and passing on those traits, those beliefs and behaviors, to their descendants, to us.

Finally, when it comes to killing and failing to respect others, we can evaluate our intuitions by assessing the ethics of killing and respect more broadly. We can consider, as we did, why death is bad for the individual who dies. In the case of death and killing, this approach provides theoretical reasons which support our intuitions. In contrast, the badness of pain and suffering, apart from the consequences, depends simply on how bad it is for the individual. There are no further aspects to assess to see whether they support our intuitions. We are left, then, without any way to check whether our intuitions are being influenced by the implicit assumption that animal suffering is not as bad as human suffering.

One day scientists might identify the regions of the brain which are responsible for these assumptions. If they do, and we can turn them off, we might be able to assess, free of these biases, whether pain and suffering really are worse in human beings. Given our current inability to do that, examples of pain and suffering, like *Experimentation*, and our intuitions in response to them, do not provide a means to evaluate whether there are degrees of moral status. Instead, to determine whether it is worse to cause humans to suffer, we first need to determine whether human beings are morally more important than animals and then apply the results of that evaluation to cases of pain and suffering. The final conclusion of this book—there are no degrees of moral status—thus provides the answer to the title question of the present chapter: *No*, pain and suffering are not really worse in humans than in animals.

6

Does Respect Really Apply Only to Humans?

Introduction

The conclusion that there are no properties which increase the moral status of those who possess them, hence, no degrees of moral status, led us to reconsider our intuitions in response to three prominent examples of significant harms/wrongs. We first reconsidered whether it is significantly worse to kill human beings than it is to kill rabbits. We have seen that it is, but this difference does not trace to a difference in degrees of moral status. We then reconsidered whether pain and suffering are much worse when they occur in human beings rather than in animals. We saw that there are good reasons to doubt the reliability of our intuitions in these cases, suggesting that they do not provide reason to question the present conclusion. This leaves our intuitions with respect to the third significant harm/wrong: Does respect really apply to humans but not animals? Two senses of respect seem to suggest that it does.

The first sense involves respect for *individuals as individuals.* While this sense of respect is widely discussed and frequently endorsed, there is significant debate over what exactly it involves. A common understanding is in terms of individuals placing limits on the extent to which they can be harmed in order to benefit others. Humans place strong moral limits in this regard. For example, it is unethical to painlessly sacrifice one human being to save the lives of five or even five hundred others. In contrast, it is widely thought that painlessly sacrificing individual animals, as

Life Without Degrees of Moral Status. David S. Wendler, Oxford University Press.
DOI: 10.1093/oso/9780197675328.003.0007

happens in medical experiments all the time, is ethically acceptable. The standard explanation for this difference in that individual human beings have high or full moral status, whereas individual animals don't, hence, respect for individuals as individuals applies to us but not to them.

The second sense of respect that we will consider involves respect for *individuals' agency*. To take a specific example, rabbits and human beings both attempt to mate and raise their offspring. But the ethics of interfering with these aspects of their lives seem very different. We routinely interfere with the ability of animals to mate and raise their offspring. We operate on them so that they cannot impregnate others or get pregnant. We take their offspring away and give them to others. When we do these things to rabbits (and cats and dogs and horses), it seems, to most people at least, routine and largely unobjectionable. In contrast, doing these things to other human beings would be seriously problematic. The standard explanation in this case is that human beings, but not animals, have a right to decide how their lives go, which includes a right to determine whether and with whom they mate, and whether they have children. And the reason why we have this right and animals don't, the argument goes, is that we are morally much more important than they are. The present chapter evaluates whether this is right, whether the ethics of respect in these two senses suggests, contrary to the present conclusion, that there are degrees of moral status.

Some commentators maintain that respect for agency applies to individuals who are autonomous in the sense of having the capacity of competent adults to direct the course of their own lives based on their views of the type of life they want to lead, and to no one else. This view, which has been very influential, is reflected in one of the most important US court cases to deal with this question, *Tarlow v. DC*. In it, the US Federal Court of Appeals argued that medical providers do not need to solicit the preferences of individuals who have lifelong decisional impairments before performing medical procedures on them, including elective surgery.[1] The court ruled that, because these individuals had never been competent, their wishes were irrelevant to whether the surgery ought to be performed. Trying to do otherwise, trying to determine and take into account the individuals' wishes, the Court argued, lacks "logical sense."

This ruling ignores the fact that the superior cognitive capacities which underlie individuals' ability to make decisions regarding how their lives go come in degrees. Moreover, individuals whose capacities to make decisions fall below those of competent adults in some respects might be equivalent to the capacities of competent adults in other respects. For example, individuals who are not able to understand probabilities and make decisions in light of them lack decisional capacity, and they are not competent. Nonetheless, they might still have a rich sense for what constitutes a flourishing life. The fact that these individuals have moral status equal to that of a competent adult implies that their views regarding what constitutes a flourishing life merit equal respect, even if they require the help of others to realize them.

Other humans have a less rich sense of what constitutes a flourishing life compared to competent adults. But the claim that their views do not deserve respect at all assumes there are morally relevant thresholds on the underlying cognitive capacities; individuals whose capacities fall below the threshold do not deserve any respect and individuals whose capacities exceed the threshold are owed respect as autonomous individuals. This would make sense if it were the case that having the necessary capacities beyond the threshold were necessary for having moral status at all. However, the possibility that sentience is sufficient for having moral status (which we will pursue in section 6.3) suggests that this view is mistaken. Moreover, as we saw in section 3.5, such thresholds do not exist, leaving us with the challenge, to be discussed in the present chapter, of determining how to respect individuals whose capacities fall below ours, including animals. Because the all-or-nothing view has had significant influence on us and our beliefs, this won't be easy, and we will need to accept the possibility that the answers to some of these questions might not be immediately apparent, and the answers to others might elude us for a long time to come.

6.1. Do Animals Deserve Respect as Individuals?

Imagine there are five people who need an organ transplant to stay alive: two need a kidney, one needs a heart, one needs a liver, and one needs a lung. This scenario raises a standard ethics question: Is it ethical to painlessly kill one healthy person in order to use their organs to save the lives of five others? Doing so would involve helping more people than it hurts, saving, in aggregate, four lives. Despite this, doing so would be unethical because killing the one person involves treating them as a means to saving the lives of the others and not as an individual who is worthy of respect in their own right.

The fact that this is immoral when it comes to human beings suggests that we are morally important in ways that place significant limits on the extent to which others may harm us, even when doing so benefits others. They may not intentionally kill us for our organs, even when doing so would save the lives of five others. In contrast, imagine that we give a rabbit a good life and then kill it painlessly in order to save five human beings, or even five rabbits. This seems like it could be an acceptable thing to do, suggesting that respect for individuals as individuals applies to human beings, but not to rabbits. Considerations of respect are thus often regarded as implying that we are more important morally than rabbits: respect in these cases applies to us, but not rabbits, and that must be because we are morally more important than they are.

While this argument makes sense, the conclusion from section 4.1 that painless death is not a terrible thing for rabbits, for their own sakes, suggests a different explanation: killing a rabbit painlessly to save others is ethically acceptable because it is not bad for the rabbit itself. If that is right, if painlessly killing rabbits does not significantly harm them, cases of sacrificing one rabbit to save five others do not provide a reliable test of whether respect for individuals as individuals applies to rabbits. For that, we need a case of treating rabbits in ways that clearly and significantly harm them for the benefit of others, for example:

> There are 20 million people with type 1 diabetes worldwide. These people die, on average, 15 years early. The development of an effective treatment or a cure which resulted in each of these individuals having an average life span would thus save many, many years of life.[2]

This example illustrates the fact that medical research can offer tremendous potential benefits. Hence, the claim that respect for individuals as individuals does not apply to animals suggests that it could be acceptable to expose a small number of animals to dramatic suffering in order to develop a possible treatment for diseases like type 1 diabetes.

To make this possibility more concrete, consider whether it would be ethically acceptable for researchers to expose five rabbits to excruciating pain for ten years continuously as part of an experiment that has a 1 in 10,000 chance of curing type 1 diabetes. If respect for individuals as individuals does not apply to animals, this could be acceptable. Put generally: if respect for individuals as individuals does not apply to animals then, in principle, there is no upper limit on the amount of excruciating pain to which it can be ethical to expose them. As long as the potential benefits for others are large enough, it can be acceptable to expose rabbits to excruciating pain continuously for ten years. Or expose giant tortoises to excruciating pain continuously for one hundred years.

But this would be ethically problematic. Precisely when treating animals in this way becomes ethically problematic is not clear.

Would excruciating pain for one hour be acceptable? For one week? For one month? While the answers to those questions are not clear, it is clear there is a limit. Exposing animals to ten years, much less one hundred years, of continuous excruciating pain for a very small chance of benefiting even millions of others is unethical. That is enough to show that respect for individuals as individuals applies to animals, at least in some cases, and to some extent. Even when the potential benefits significantly outweigh the harms, some ways of harming animals are unethical.

To consider a second example, imagine yourself back on the committee mentioned in *Experimentation*, which is now reviewing two possible studies:

> Experiment 1: Continuous excruciating pain to one rabbit for one month
>
> Experiment 2: Moderate pain for a very brief time to each of one hundred rabbits

Imagine that the total potential benefits and harms are the same in the two cases. The only difference is whether all the pain and suffering are experienced by one rabbit, in which case that rabbit is subjected to a dramatic amount of pain and suffering, or the pain and suffering are distributed across one hundred rabbits, in which case each rabbit experiences moderate pain for a very brief time.

If respect for individuals as individuals is irrelevant to animal ethics, there would be no moral reason to prefer one version over the other. Morally speaking, you could pick either one since the total harms and benefits are the same. Still, ethically speaking, the second version seems better. It is better ethically to distribute the suffering across one hundred rabbits rather than locating it all in one rabbit. This conclusion provides additional evidence that respect for individuals as individuals applies to animals, at least in some cases and to some extent.

A third example: Standard practice involves testing experimental treatments in one or two animal models before testing them in human beings. This approach allows researchers to evaluate the toxicity of experimental treatments with animals, with the goal of screening out especially toxic ones, before testing them in human beings. Imagine, then, an investigator who is studying a potential new treatment for cancer in human beings. Following successful studies in the laboratory, she plans to inject it into a single rabbit and monitor the rabbit for any signs of acute toxicity. Because many experiments are not successful the first time, the researcher purchases two rabbits just in case.

She injects the first rabbit with the experimental treatment and draws its blood a few times over the next forty-eight hours. This time, the experiment works the first time. The first rabbit experiences significant discomfort for a few hours, but shows no signs of an acute allergic reaction, in which case the researcher does not need to inject the second rabbit. Just as the researcher is about to release both rabbits, she finds a treat she can give to one of them before they go. Which rabbit should get the treat? If respect for individuals as individuals does not apply to animals, she should maximize the total amount of good experiences. Hence, if both rabbits will enjoy the treat equally, it doesn't matter ethically which one receives it. The researcher can pick the rabbit she prefers or she can flip a coin. Ethically, that would be a mistake. The researcher cannot undo the pain caused by the experiment. She also cannot increase the amount of pleasure that will be experienced by whichever rabbit eats the treat. But the researcher can make it the case that the pleasure of eating the treat is experienced by the rabbit that was subjected to the pain of the experiment. That provides an ethical reason to give the treat to that rabbit, suggesting respect for individuals as individuals applies to animals in this case as well.

Taken together, these three examples suggest that the claim we started with—respect for individuals as individuals applies to humans, but not animals—is mistaken. It applies to both humans and animals. This is not to say that respect for individuals as individuals applies equally to humans and animals, nor that it applies equally to different animals. That depends on the basis of respect in this sense and whether the basis or bases are equally relevant across different types of individuals. We will pursue this question in section 6.3. The conclusion here is simply that respect for individuals as individuals applies to sentient animals, which reveals that, to this extent, the ethics of respect does not provide a reason to think there are degrees of moral status. Now consider respect for individuals' agency. Does respect for agency apply to human beings but not animals? And, if so, does this difference provide reason to think there are degrees of moral status among those who matter morally?

6.2. Should Animals Decide How Their Lives Go?

There are different ways to understand agency. To begin, we can think of it very broadly as self-directed movement, or moving through one's environment based on a personal reason or with a personal goal.[3] Respect for agency in this sense involves allowing others to do what they attempt or choose to do. Failure to respect others in this regard involves disrupting, interfering, getting in the way, or stopping them from doing what they otherwise would attempt or choose to do.

Whether we respect or interfere with individuals' agency frequently affects their experiences. Stopping two-year-olds from doing what they want to do often has a negative impact on how they feel, as evidenced by their kicking and screaming. The fact that the experience is upsetting provides a *welfare-based* reason to allow two-year-olds to do what they want. The same consideration applies to animals. To the extent that interfering with their lives results in their having negative experiences, we have welfare-based reasons to refrain. To that extent, we are not respecting their agency, however; we are looking out for their interests or their welfare.

When it comes to adult humans, respect for agency is not limited to concern for their welfare. Even if I would be happier if you chose my spouse, or what I have for lunch, respect for my agency provides reason for you to allow me to decide. To assess whether respect for agency supports degrees of moral status, we can consider whether similar considerations apply to animals. To do that, we can consider what relationship must obtain between an individual and the individual's actions for it to be the case that there is a moral reason to respect their agency for the sake of the individual themselves.

The standard answer, very briefly, is that there are non-welfare-related reasons to respect the agency of individuals when they are acting autonomously. This view is reflected in the *Tarlow v. DC* case mentioned in the introduction. While there are a wide range of views regarding what acting autonomously amounts to, we can understand respect for agency in this sense as applying to individuals who have a rich view of what type of life is best for them, understand the potential benefits and the risks of the available options, and have the capacity to make a choice based on these things.

If respect for agency were limited to individuals who are autonomous in this sense, there would be no reason to respect the agency of animals. In particular, most animals do not have a rich view of what type of life is best for them. This characterization also implies that respect for individual agency is irrelevant to human beings who are not autonomous in this sense. Most prominently, many adults with moderate dementia do not possess the capacities necessary for autonomous action. Nonetheless, their preferences and choices have moral weight, which provides non-welfare-related reasons to take them into account when deciding how to treat them.

This view is endorsed by most guidelines and regulations for medical research. For example, the Declaration of Helsinki mandates that investigators solicit the views and obtain the agreement of adults who have lost the ability to provide their own informed consent before enrolling them in research.[4] Respect for agency also applies to many adults who have never been autonomous. A number of years ago, I was asked to evaluate a woman with a rare neurological disease which resulted in her having cognitive capacities similar to an average six-year-old. She could understand basic facts, but she had trouble understanding abstract ideas and had, at best, an unclear understanding of the future. The disease ran in her family and the investigators were hoping to obtain a biopsy to learn more about it, with the goal of trying to find its cause and hopefully use that information to develop a treatment. The potential benefits of these developments were significant, but the chances of realizing them were unknown and likely low and, if they were realized at all, it would not be for years.

The woman understood that the study would require her to undergo a procedure that would cause her some discomfort and pain. She also understood that the procedure would not help her medically, but it might help the researchers learn things that could help other people with the same condition, including her family members. In contrast, she could not understand the long-term risks, and she did not understand how information collected from her might someday lead to a treatment for her condition. Still, based on what she did understand, she very much wanted to participate in the study.

This desire to participate and potentially to help others was consistent with her view of how she wanted her life to go. Her life was oriented to doing things to help others, and pursuing such activities gave her life meaning. Because she did not understand as well as competent adults, her preferences did not have the same moral weight as theirs. But she did understand essential aspects of the procedure, and based on that understanding, she very much wanted to participate. Her preference to enroll thus provided an important moral reason to enroll her.[5] Doing so gave her the opportunity to pursue her own view of a flourishing life, rather than simply being the passive recipient of the decisions made by others.

This example reveals that respect for individual agency is not limited to autonomous individuals, to what some philosophers refer to as *persons*. It applies to at least some human beings who are not autonomous. And this conclusion raises the possibility that respect for agency is not limited to human beings. If there are animals whose capacities are similar to this woman's, respect for individual agency would apply to them as well. This leaves us with the challenge of determining what conditions individuals must satisfy for respect for individual agency to apply to them, and whether some animals realize those conditions.

There is significant debate in this regard. Is having preferences or values sufficient to merit respect for one's agency? Or does it require that one has the capacity to understand one's options? To be able to reason about the different options? To choose between the options based on one's own view of what constitutes a flourishing life? Fortunately, we do not need to answer these questions. Instead, the important point is that these underlying capacities come in degrees, which define a spectrum of levels of agency or agents. Some individuals are very simple agents, and others are increasingly sophisticated ones. Some have limited control over their actions (e.g., infants), and others have greater control (e.g., competent adults). Some have limited understanding of the options, and others have more. Some have very simple preferences with little or no capacity to reflect on and change their preferences. Others are more sophisticated in this regard.

Proponents of degrees of moral status maintain that respect for agency applies to individuals who are autonomous and not to anyone else. However, the conclusion of section 3.5 that there are no morally relevant thresholds on properties like being autonomous undermines this view. It suggests instead that respect for agency comes in degrees which track the level of the agency of the individuals in question and the extent to which directing the course of their own lives is part of a flourishing life for them. Once we reject degrees of moral status and regard individuals' agency as moral action guiding, rather than moral status enhancing, this makes sense. Instead of seeing autonomy and respect for agency as all or nothing, we see it as a matter of degree, which explains why the preferences of the woman I evaluated had some moral significance, despite the fact that she was not autonomous.

This analysis is not sufficient to determine which animals are covered by respect for agency, and to what extent. It does not determine whether we should respect the agency of mice, rabbits, or alligators. Making those determinations will require a more in-depth assessment of precisely which capacities are needed for there to be an ethical reason to respect an individual's agency, and which animals possess those capacities. This is important work to do. However, it is not necessary for present purposes. The question we face is whether the ethics of respect for agency applies to human beings and does not apply to animals.

We have seen that agency comes in degrees, and there is correspondingly greater reason to respect the agency of more sophisticated agents. The fact that we are fairly sophisticated agents explains why it is important to respect our agency. However, it is not the case that there are no non-welfare-based reasons to respect the agency of individuals who are not autonomous to the same extent. This reveals that respect for agency is not limited to human beings who are autonomous to the extent of competent human adults, and it might apply to some animals as well. Consideration of the ethics of respect for agency thus does not provide reason to think that there are degrees of moral status. This conclusion raises the question, to be considered in the next section (section 6.3), of what provides the foundation or the basis of the ethics of respect. Why, in other words, does respect for individuals as individuals apply to animals, and why isn't respect for agency limited to autonomous individuals? First though, it will be important to consider what the present conclusion does and does not show.

We have been assessing the extent to which there are reasons to respect the agency of individuals, for the sake of the individuals themselves. While this evaluation is important for determining whether there are degrees of moral status, it considers only one reason to respect individuals' agency. Moral status is not the only value in the world, and respect for individuals, for their own sake, is not the only consideration relevant to determining how they ought to be treated. In section I.3, we considered works of art, like the Mona Lisa, and wonders of nature, like Victoria Falls. They do not have moral status, but it would still be problematic to destroy them. The reason is that they have other types of value which merit respect.

To take a different example, the Venus flytrap closes in approximately 100 milliseconds and the Prayer Plant moves its leaves into a vertical ("praying") position in the evening. These plants do not have moral status; hence, respecting these movements is not a matter of treating the plants appropriately for their own sake. Nonetheless, these plants and their movements merit respect as amazing adaptations to the plants' environments. Similarly, the movements of animals might merit respect even if this is not a matter of respecting their agency, but a matter of respecting them as amazing and complex beings. To the extent that we are concerned with determining how we ought to treat animals, we should not limit ourselves to respect for individual agency. We should also consider whether they merit respect in other ways, and what types of treatment are consistent and not consistent with respecting them in those ways.

6.3. The Connection Between Sentience and Respect

Proponents of degrees of moral status believe that respect for individuals as individuals and respect for individuals' agency apply to average human beings, but not to animals. They explain this difference by endorsing complex theories which maintain that respect in these senses applies to individuals who possess certain superior cognitive capacities. It applies, for example, to individuals who are moral agents, or autonomous, or who have a narrative self-identity. Hence, while sentient animals have moral status, they have lower moral status and do not have claims to be treated with respect. We have seen that this view is mistaken. This conclusion thereby presses the question of why respect is not limited to persons or human beings. If possession of one or more superior cognitive capacities does not provide the basis for meriting respect, what does?

As we saw in section 6.1, it is better to distribute a significant amount of suffering across one hundred rabbits, so that each one experiences only some suffering, rather than locating it in one rabbit, which suffers extensively. Similarly, the fact that one rabbit was exposed to pain and suffering in an experiment provides a reason to give it a treat as opposed to giving the treat to another rabbit, even when the second rabbit would enjoy it as much. These conclusions were based, not on whether the rabbits possess some superior cognitive capacities, but on the fact that they are sentient: they suffer when they experience pain, and they experience pleasure when they receive treats. This suggests the possibility that sentience provides the basis for meriting respect. Pursuing this possibility provides a way to investigate the present conclusion that there are no degrees of moral status among those who matter morally. Put simply: why is sentience so important morally? Why is it, rather than moral agency or autonomy, the property which endows individuals with moral status?

The fact that some individuals are sentient changes the world in a fundamental way. It implies that there is more than just stuff in the world. There are subjective experiences and the individuals who have them. There are subjective worlds, which include sensations, thoughts, pain, and pleasures, that depend on the individual's existence and to which only one individual, that individual, has access. Only they experience the pain or anxiety or depression in their world, and only they know what those experiences are like. The very existence of that world depends on their existence. This helps to explain why sentience is moral status conferring.

Moral status is about how we treat individuals for their own sakes. It implies that they matter morally, and there are moral reasons to avoid causing them to have aversive experiences. Consider the following passage:

> Respecting others' decisions is not simply a way to promote another's welfare or to facilitate the valuable process of creating and expressing a distinctive character. In a more basic way, I believe it serves as an acknowledgment of the moral importance of the uniqueness and separateness of persons and the deep, irreducible fact that one's life is the only life one has.[6]

This passage is intended to explain why it is problematic to interfere with the choices of *autonomous* individuals. But it suggests a broader point. It suggests that there are moral reasons to respect all sentient agents as a way of respecting their subjective world and their subjective lives. It is not so much, as the author writes, that one's life is the only life that one has. That is true, but, as we have seen, ending one's life, death, is not nearly as bad for us as we are inclined to think. Instead, the important point is that there is only one individual who has that life. It is a unique world that depends on their existence and to which only they have direct access.

We talk about trying to understand the feelings of others. Their anger, their sadness, their joy. And we can try to imagine what it is like for them, how it feels. But, in a fundamental sense, we can never succeed. We can never have the experiences of others, and we can never know what it is like to live in their world, to be them. Theirs is a unique world, and only they experience it.[7] This explains why the things mentioned in section 1.3, Victoria Falls, Machu Picchu, the Mona Lisa, and the Magna Carta, don't have moral status, even though they have great importance and value. Destroying these things would be terrible. But these things themselves aren't sentient; hence, they don't have a personal stake in whether they are destroyed or not.

What is involved in respecting these unique worlds depends on the nature of the existence in question. Projects and relationships take time. Hence, individuals who truly live in the present moment cannot pursue them. For them, life might simply be a succession of momentary and unconnected feelings of pain and pleasure. In that case, respect involves not causing them to suffer and helping them to experience pleasure. But the fact that sentience is the basis for moral status does not imply that pain and pleasure are all that matter morally. Instead, sentience is so important morally because it gives individuals a stake in how their lives go. At that point, we need to consider what constitutes a better or flourishing life for the individual, for their own sake.

While a great deal is unknown, we are learning that, far from being machines, the subjective lives of animals are much richer and more complicated than was previously recognized. It appears that animals, including chimpanzees, crows, elephants, and wolves, form goals and act on the basis of them.[8] Animals also coordinate their behavior to attain commonly shared goals. To briefly consider a fascinating example, many animals live in social groups, and the coherence of the group requires that they coordinate their behavior. They need to decide when the group stays put and when it moves on. Different species have evolved different strategies for making these decisions.[9] For some, each member expresses their opinion.[10] African Wild Dogs decide whether to go out on a hunt by making a kind of sneezing noise. While the leader's preference counts for more, the preferences of all the members count.[11] Similarly, many animals are cognizant of and respond to the preferences and needs of their offspring.[12] These behaviors display a capacity for acting based on the individuals' own perspective on the world.

One might worry that the challenge of thresholds arises for this view of moral status as well. Sentience or awareness comes in degrees. Some animals might have only the most rudimentary sentience; their connection to the world of experience might be similar to our sense of the world just before we are knocked out by the anesthesia. From there, there are steadily increasing levels of sentience and awareness. If having moral status depends on being sentient, we will have to identify a threshold on this spectrum where individuals go from having no moral status to having moral status. At that point, the previous challenge recurs: isn't any choice of a threshold arbitrary in the sense that there is no justification for treating individuals whose sentience falls just below the threshold as having no moral status and those just above it as having moral status?

The answer in this case is no. Being sentient to any extent endows individuals with moral status; they don't have to be sufficiently sentient. However, as noted earlier, the fact that an individual has moral status implies only that how they are treated matters morally. It does not tell us how they ought to be treated. For that, we need to look to their moral action guiding properties and what they have at stake. Since these come in degrees, the strength of individuals' claims to not be harmed comes in degrees. The greater an individual's sentience, the more that they have at stake in many contexts; hence, the greater the reasons to treat them well and to avoid treating them ill.

At a minimum, the present discussion undermines the claim that there are important moral reasons to respect human beings or persons, but no moral reasons to respect sentient animals. Identifying the implications of this conclusion for whether and to what extent there are moral reasons to respect the agency of animals requires a determination of when and to what extent directing the course of one's life is important for one's own sake. Is it sufficient to have a plan for how one's life goes? Or does one need to be aware of or even care about the realization of the plan? What gives the plan value: the fact that it is aimed at? Or that one knows one is aiming at it? Or that one knows that and values realizing it?[13] Another option: the fact that the individual is an independent source of direction of its life and attempts to direct the course of its own life provides some reason to respect that capacity. It provides a way to respect the subjective life of that individual.

These reasons are moral to the extent that respecting the individual's agency is important for the sake of respecting the individual themselves for their own sake. As mentioned in section 6.2, there might be other reasons to respect the agency of some individuals as well. It might be that we should respect the agency of some animals because they represent an amazing adaptation to their environment over millions of years. That may deserve respect in the way that trees deserve respect, even though they do not have moral status. But the fact that there can be moral reasons to respect the agency of individuals who are not autonomous undermines the claim that respect applies only to competent human beings and, thereby, undermines an important argument for degrees of moral status. To that extent, these considerations support the present conclusion that there are no degrees of moral status.

Summary

In the present chapter, we considered whether two types of respect really apply to human beings, but not animals. We have seen that respect for individuals as individuals applies to animals, and respect for agency is not limited to autonomous individuals (persons) and may well apply to at least some animals, to some extent. These findings, combined with the findings of Chapters 4 and 5, reveal that the ethics of three of the most significant ways to treat others negatively does not undermine the present conclusion that there are no degrees of moral status among those who matter morally.

Proponents might accept this conclusion but argue that differences in degrees of moral status are relevant to other ways of treating individuals. While that's possible, the fact that these three ways of treating others are so important makes it unlikely.

If differences in degrees of moral status are not relevant to how we treat others in these three ways, there are no degrees of moral status. Chapter 8 thus considers the implications of this conclusion. What should we make of the fact that we live in a world without degrees of moral status? What does this conclusion imply for how we ought to live and how we ought to treat others? Before taking on those questions, Chapter 7 considers two arguments which accept the present conclusion but maintain that we should believe in degrees of moral status anyway.

7
Are there Other Reasons to Believe?

Introduction

Degrees of moral status depend on some individuals possessing a property or properties, moral status enhancing properties, which increase their moral status relative to other individuals. We have considered the conditions that would have to exist for some properties to be moral status enhancing and concluded that no properties have this effect. This conclusion conflicts with the intuitions on which endorsement of degrees of moral status is typically based. We thus reconsidered these intuitions in more depth to determine whether it is possible to account for them without appeal to degrees of moral status. The fact that it is possible further supports the conclusion that there are no degrees of moral status. Granting this conclusion, the present chapter considers two arguments which claim that we should, nonetheless, continue to believe that we have higher moral status than animals.

7.1. Degrees of Moral Status as a Rule of Thumb

Human beings possess many moral action guiding properties which animals lack or possess to a lesser extent. In many circumstances, these differences imply that it is morally worse to treat us negatively. It is, for example, worse to kill human beings

Life Without Degrees of Moral Status. David S. Wendler, Oxford University Press.
DOI: 10.1093/oso/9780197675328.003.0008

than rabbits, worse to interfere with our agency. The fact that frequently it is worse to harm and fail to respect human beings offers a possible explanation for why many people believe there are degrees of moral status: the fact that it is *frequently* more important to treat human beings well is thought to imply that morally we are more important than they are.

We have seen that this is a mistake. It is more important to treat human beings better in many cases, not because we are morally more important, but because our superior cognitive capacities imply that we often have more at stake. This suggests a different way to think about whether we should believe in degrees of moral status. The right thing to do, in many cases, is to treat human beings better than animals. Furthermore, believing that we are morally more important increases the chances that we will favor human beings in our actions. This suggests that believing we are morally more important than animals might increase the chances we do the right thing, providing reason to believe there are degrees of moral status, even though we have seen that there aren't.

This pragmatic argument for believing in degrees of moral status gains some plausibility from the fact that it can be difficult to determine how much a given action benefits or harms human beings compared to animals. Exactly how will being in the experiment affect animals, what will be the long-term consequences for them, and how do they compare to the impact on human subjects? Answering these questions and, thereby, determining who has more at stake, can take a good deal of time, which has its own costs. In addition, going through the calculations might increase the chances we get things wrong. To appreciate these challenges, imagine we teach our children that there are no degrees of moral status. Instead, to determine what morally they ought to do, we teach them to determine who might be affected by their actions and what those individuals have at stake. Someone who was so trained now faces the dilemma described in *House on Fire*: they are standing in front of a burning house which is occupied by one rabbit and one person, with enough time to save one of them, but not both. How would someone who was raised to recognize that there are no degrees of moral status react? Here is one possibility:

> Which individual I should save depends on how much will be lost with the death of each one. I should add up what is lost if the rabbit dies, add up what is lost if the human dies, and compare them. The richness of the life that is lost in the case of the human being depends on the extent to which they have superior cognitive capacities. Maybe they don't. Maybe this human being had a terrible stroke which eliminated their superior cognitive capacities. I could go into the house and quiz them, but that might take a while and sounds dangerous. Okay, that's a bad idea. I could quickly check on my phone to see what percentage of human beings have had strokes and whether I am in a neighborhood where such strokes are more or less likely . . .

Completing this calculation for all the relevant considerations would take a very long time. In the meantime, both the rabbit and the human being would die. This suggests that recognizing there are no degrees of moral status, and acting accordingly, might lead to morally worse outcomes. It might lead to our frequently doing the wrong thing in the process of trying to do what is right.

One way to avoid this concern would be to teach people, when faced with these dilemmas, to flip a coin. Heads, I save the person; tails, I save the rabbit. The problem is that, in most cases, it is better to save the human being. A better response, then, one which maximizes the chances of doing the right thing, would be to save the human being every time. And, arguably, the most reliable way to get people to do that is to teach them that human beings have higher moral status than animals.

This suggests that the pragmatic argument for believing in degrees of moral status makes sense in cases like the *House on Fire*. However, these cases were designed to elicit our intuitions regarding moral status, not to represent the kinds of decisions we typically face. To determine whether this argument provides reason to believe in degrees of moral status, then, we need to go beyond these examples and consider its consequences more generally. While going through all the options, and considering the consequences of our believing in degrees of moral status for each of them would require several books, history provides a shortcut.

For thousands of years, most human beings have believed that we are morally more important than animals. And what have been the results? It certainly has led to humans privileging ourselves over animals. But has it led to a morally better world? To humans doing the right thing more often than we would have otherwise? It hasn't. The view that we are morally more important than animals has not led simply to our favoring ourselves in tradeoff cases like *House on Fire*, where there is a good deal at stake for both some human beings and some animals, and it is not possible to benefit both. Instead, the view that we are significantly more important morally has frequently led to humans treating animals horribly, often in cases where human beings had little to gain. We have raised animals in ways that cause them to suffer terribly because treating them more humanely would have increased the cost of dinner; we have subjected them to painful experiments to develop new shades of lipstick. This history provides perhaps the best possible estimate for the implications of our believing that we have higher moral status than animals. And it suggests that this belief leads to the view that animals have very little, if any moral status, which leads to significantly unethical behavior, to our treating them badly in order to realize minor benefits for ourselves.

Advocates might respond that the problem here is not the belief that there are degrees of moral status; it is the belief that we are *significantly* more important morally than animals. If we believe that human beings are only slightly more important morally, the problems with abusing animals would be less likely to occur. While this is possible, it also undermines the value of adopting the rule of thumb in the first place. The belief that we are only slightly more important implies that we should benefit human beings when all other relevant considerations are equal, or at least close to it. But the point of believing in degrees of moral status is to avoid having to go through the process of assessing whether all else is roughly equal. If we have to do that anyway, the heuristic loses much of its value and what value remains does not seem to justify teaching our children to believe things we know to be false.

Finally, endorsing the truth not only tells individuals how they should act, it explains why they should act in those ways. That understanding seems valuable in itself and may increase the chances that individuals act ethically. And this approach has the virtue of not depending on a campaign to discourage people from learning the truth. In the end, then, the fact that it is often ethically appropriate to treat humans better might explain why many people believe there are degrees of moral status. But, this fact does not provide reason why we should believe this.

7.2. Degrees of Moral Status as Proper Partiality

Many people believe that we should be partial to some individuals over others, to our children, our siblings and our parents. This raises the question of whether the belief that we are morally more important than animals represents another instance of proper partiality, not as reflecting the relative importance of humans versus animals, but reflecting the fact that we should be partial toward fellow human beings. The philosopher Tim Scanlon, for example, argues that the relationship we have to other human beings provides reason for us to treat them as having higher moral status.[1] The claim here is not that we have higher moral status than rabbits, but that we have reason to favor human beings over rabbits because they are our fellows.[2]

Partiality *within* human beings provides some support for this view. Human beings are equally important morally. However, it does not follow that morally we ought to treat all human beings the same. I should do more for my siblings than I do for your siblings, and you should do more for yours than you do for mine. It is not simply that we happen to care more for our own siblings; morally we ought to favor our own siblings. If the *House on Fire* holds your brother and my brother, Martians would have no moral reason to prefer one over the other. They perhaps should flip a coin. But I am not a Martian. I am my brother's brother, and that gives me a moral reason to rescue him, just as it gives you, in the same situation, a moral reason to rescue your brother.[3] Does proper partiality also give us a moral reason to favor other human beings over animals? Some people think so. Baruch Brody argues that our favoring human beings over animals "represents one more example of discounting the interests of strangers, a feature that is pervasive in morality."[4]

Other people think that discounting the interests of strangers compared to those we know and love is inappropriate, unethical. If they are right, then proper partiality does not provide a reason to believe in degrees of moral status. Granting that possibility, it does not seem unreasonable to think that partiality within humans is appropriate in some cases, providing reason to consider whether it might support the claim that we should regard human beings as being more important morally.

The literature on partiality attempts to explain why it can be appropriate to treat some human beings (e.g., our children, our siblings) better than others, even though all human beings have equal moral status.[5] This work highlights three important challenges facing the claim that proper partiality supports our favoring human beings over animals. First, proponents argue that a system of privileging those closest to us yields important benefits that would not be available without it. To take one example, the benefits that result from being a member of a family depend on family members being partial to each other.[6] My parents treating all children equally, considering them all equally important to them, would be incompatible with my deriving the benefits of having parents who are dedicated to me in particular. This suggests that it can be acceptable for my parents to privilege me as long as other children have parents who privilege them. In other words, the acceptability of our treating some individuals better depends on everyone having similar relationships. Proper partiality does not justify parents devoting all their resources to their own children and contributing nothing to the care of orphans who have no one to privilege them.

This aspect of proper partiality raises concern when it comes to animals. At a minimum, it suggests that we can justify favoring other human beings only to the extent that animals have the capacity and resources to look out for each other. The problem, of course, is that many animals do not have the capacity to protect and benefit their own anywhere near to the extent that human beings do. They do not have the capacity to protect their children from diseases, from predators. Hence, even if we endorse partiality within human beings, it does not seem to justify our favoring other humans and ignoring the needs of animals. Put differently, once we recognize that animals count as much morally as humans, we have to take into account not just the impact our partiality for other humans has on us; we have to take into account the impact it has on animals.

This raises a second challenge. Proper partiality depends on preferring and privileging some individuals over others. However, proper partiality itself does not determine who we ought to privilege. The claim that proper partiality supports a belief in degrees of moral status assumes that we should draw the lines of partiality based on the species to which we belong. Of course, human beings are more closely related to other human beings in evolutionary and genetic terms. But human beings are many other things beyond *Homo sapiens*. We are mammals, we do not have tails, we are born in a particular part of the world, and we are living things. This raises the challenge of determining to whom we should be partial. Maybe I should favor all individuals who were born in New Jersey independent of the species to which they belong.

Clare Palmer argues that we have obligations to assist human strangers with whom we have no personal relationships because of the richly related community that we share with them. In contrast, we do not have the same rich relationships with wild animals: "This is the case irrespective of whether assistance would be good or bad for them as individuals and independent of any other concerns about the consequences of assistance."[7] Palmer cites work by Francis and Norman who argue that only human relations are sufficiently rich and complex to constitute a morally significant community.[8] At the same time, she points out that some human beings have relatively impoverished relations with other human beings, and some domesticated animals have deep relationships with human beings. Many people consider their pets to be important members of their family and frequently prioritize taking care of their pets and spending time and resources on them over other human beings. Palmer concludes that Francis's and Norman's view should be expanded to a global social community which includes all humans and sentient animals with whom humans have strong social relations.[9] This view seems to suggest that proper partiality depends on one's preferences and relationships. People who have significant relationships with horses should privilege horses, birders should privilege birds, cat lovers should favor cats, humanists should favor other humans, and perhaps curmudgeons should privilege only themselves. This approach thus does not support the claim that human beings in general should privilege other human beings over all animals.

Bernard Williams argues that a preference for human beings, what he calls the human prejudice, is a fundamental aspect of our view of the world and our way of life.[10] In supporting other human beings over animals, we are, in effect, defending our culture or our way of life. To determine whether this view supports favoring humans over animals, we need to determine what, in fact, is our way of life. Centuries ago, defending human culture involved defending the enslaving of one's enemy, the ownership of one's wife and children, and the abuse of animals. Today, it involves according to women equal rights, and recognizing that animals matter morally. To this, animal rights proponents argue that we should add equal rights for animals. We can't reject this claim on the grounds that it is inconsistent with human culture since it involves a claim about what human culture is and should be.

Proponents might respond that a preference for our species over other animals is a fairly fixed fact about human psychology. It might be. But it also might be the case that we are developing toward a culture, a way of life, which recognizes that animals have equal moral status.[11] This possibility is suggested by the fact that there are many people who currently think there are no degrees of moral status. And there are people who care for certain animals more than other human beings, people who devote their time and money helping whales because they care about them more than they care about their neighbors. I don't see any reason to think these people are betraying the human way of life as opposed to living it.

The third problem: even if one accepts the moral appropriateness of partiality, it is limited in ways that degrees of moral status are not. In particular, proper partiality does not justify harming others to benefit those to whom we are partial. Imagine my sister is very ill as the result of a rare disease. Partiality allows me to spend my money helping her get treatment. It allows me to spend my time caring for her. But proper partiality does not allow me to take food from other people to benefit my sister. It also does not allow me to conduct experiments on them for the sake of my sister. Imagine my sister's medical team determines that her illness traces to her kidneys. They tell me that, if they had kidney tissue from five other women of a similar age, they could use that tissue to test possible treatments for my sister. Proper partiality does not justify my kidnapping five strangers and performing invasive surgery on them to obtain pieces of their kidneys for the researchers. It does not even justify my sneaking into the rooms of five female patients and taking a little of their blood while they sleep. This is true even if they won't notice it and my taking the blood poses no risks to them. In contrast, degrees of moral status are thought to justify our harming animals in precisely these ways when doing so has the potential to benefit us.

This analysis reveals that proper partiality can justify favoring some individuals over others only to the extent that everyone who matters morally is taken care of to a sufficient extent. In addition, it is not clear that proper partiality provides a reason for us to favor all other human beings over all animals. Perhaps it supports our favoring those we happen to care most about, whether humans or animals. Finally, proper partiality does not justify harming others in order to benefit those we favor. These considerations, taken together, suggest that proper partiality does not provide grounds for us to favor human beings over animals in the ways that would be appropriate if we were morally more important than animals. They thus suggest that proper partiality does not provide a reason to believe in degrees of moral status.

Summary

The present chapter considered two reasons why it might make sense for us to believe in degrees of moral status even though all individuals who have moral status have it to the same degree. First, in many situations, human beings have more at stake than animals. This raised the question of whether degrees of moral status might be a useful rule of thumb: believing that human beings are more important morally is more likely to result in our doing the right thing. We have seen that, for several reasons, this belief is more likely to result in our acting inappropriately rather than appropriately.

We then considered whether the belief that we are morally more important reflects proper partiality for fellow human beings. Partiality can be appropriate in contexts in which everyone has others who are partial to them, and who are able to help them sufficiently. While some animal species might be able to help each other in this way, many others can't. Proper partiality also does not provide a basis for determining precisely which individuals we should favor. Perhaps individuals who are attached to their pets should favor them; those who are attached to birds should favor them. It thus does not support the claim that human beings should favor each other over all animals. Finally, partiality does not justify harming others for the benefit of those to whom we are partial. It does not justify putting them in cages or using them in experiments for our benefit. In the end, then, this discussion suggests not only that there are no degrees of moral status, but we should accept this fact and focus on trying to determine what it implies for how we ought to live. That is the topic for Chapter 8.

8
A World Without Degrees of Moral Status

Introduction

Once it became clear that we inhabit a corner, rather than the center, of the physical universe, people began to wonder. If we are moving through space, could we run into something? If the universe wasn't built around human beings, could there be life in other places? What does it mean for how we should think about ourselves, for how important we are? The conclusion that we do not inhabit the center of the moral universe raises similarly important questions. The present chapter considers seven of them: (1) What's the difference between a world that includes degrees of moral status and one that doesn't? (2) Do we have to spend all our time helping animals? (3) Do animals have rights? (4) What are the ethics of raising animals and eating them? (5) What are the ethics of animal experimentation? (6) What are the implications for humans who lack superior cognitive capacities? and (7) What are the implications for superbeings, especially genetically enhanced human beings, human-animal chimeras, and robots?

8.1. What's the Difference?

The conclusion that the Earth is not the center of the universe did not immediately answer every question regarding the nature of the physical universe. It did not, for example, tell us whether

Life Without Degrees of Moral Status. David S. Wendler, Oxford University Press.
DOI: 10.1093/oso/9780197675328.003.0009

the universe is static or expanding and, if the latter, how fast it is expanding and whether it will continue to expand (questions cosmologists are working on today).[1] Instead, accepting this fact provided guidance on how to go about answering these questions. Similarly, the conclusion that there are no degrees of moral status does not itself answer all the questions it raises, including how we ought to treat animals. It does not reveal whether it is important to respect the agency of rabbits. Instead, it clarifies what is and what is not relevant to making these determinations. It clarifies that, in order to determine how to treat animals, we need to look to their moral action guiding properties, not their level of moral status. We need to consider what they have at stake: How would they be affected by different ways of treating them? We also need to consider what proper respect implies for how they ought to be treated. Like scientists' reaction to the rejection of a Ptolemaic universe, determining how we should act in a moral universe without us at its center will require living in and exploring it.

The belief that we have significantly greater moral status simplifies the assessment of how we ought to treat others. In effect, our significantly greater moral status acts as a kind of trump card which frequently outweighs all other considerations and implies that we should be treated better. Without that possibility, we need to take into account a wider range of factors when determining how to act. In many cases, the right thing to do will not differ significantly from a world that includes degrees of moral status. The reason is that, although our possession of superior cognitive capacities does not endow us with greater moral status, it frequently implies that we have more at stake than animals and, therefore, should be treated better. For example, as we saw in section 4.3, it is worse to kill human beings than it is to kill rabbits. This is not because we have higher moral status, but because more is lost with our deaths than the death of rabbits, and because we have a greater claim to direct the course of our own lives.

In other cases, the conclusion that there are no degrees of moral status significantly changes things. The fact that an individual possesses superior cognitive capacities often implies that they have more at stake. But it implies other things as well, for example, that they can give informed consent. And giving informed consent frequently makes it acceptable to treat individuals in ways that would not be acceptable otherwise.

The fact that rabbits lack superior cognitive capacities and, therefore, cannot consent implies that they are frequently less able to protect themselves compared to human beings. In the context of medical research, individuals who are less able to protect themselves are considered "vulnerable" and are accorded more protections, not fewer. In the words of the Belmont report, one of the leading documents in research ethics, humans who have diminished cognitive capacities merit greater protection as a "matter of social justice."[2] Guidelines around the world thus place strict limits on the risks that are permitted in research with children and adults who lack superior cognitive capacities (e.g., those with dementia). Because current guidelines for research with animals are based on the assumption that they have lower moral status, we will need to reconsider what guidelines and regulations are needed to ensure animal research is conducted appropriately. That is the topic for section 8.5.

As we go through these issues, it will be important to keep in mind that the present conclusion might affect not only what different moral principles imply for how we ought to act, but our understanding of the moral principles themselves. In particular, we will need to consider the possibility that our current understanding of different moral principles reflects the implications they have for human beings in particular. This would not be all that surprising. To the extent human beings have significantly higher moral status than other individuals, morality is primarily about us, and our lives. Hence, we determine the implications of a moral principle by identifying its implications for us. The conclusion that there are no degrees of moral status reveals that this is a mistake. To correct it, we will need to assess whether the implications various moral principles have for humans are the same as the implications they have for rabbits, sloths, dolphins, chimpanzees, mice, and other animals.

To consider an example to which we will return in section 8.4, it is unethical to bring human beings into the world, treat them well, and then kill them painlessly for food. That is unethical because it is inconsistent with proper respect for humans. In particular, the fact that the humans would not have existed otherwise does not justify treating them as a source of food. Does the conclusion that rabbits have the same moral status as humans imply that it is equally problematic to treat rabbits in this way? That depends on whether the principle of respect for individuals implies that it is wrong to treat any individual with moral status in this way or whether it implies that it is inappropriate to treat humans, and individuals who are relevantly like humans, in this way. To the extent that it is the latter, we will need to consider which, if any, animals are similar to us in the relevant ways.

8.2. Do We Have to Spend All Our Time Helping Animals?

We have tried to assess whether there are degrees of moral status without committing to any particular moral theory. As a result, we have focused on the ethics of treating others badly: killing them, causing them to suffer, failing to respect them. This is because essentially all moral theories direct us to avoid harming or wronging others. Not all theories likewise direct us to help others. With that said, many people think, to the extent that we do help others, we should prioritize helping those who are worse off. To see why, imagine you have one treat and two children who would enjoy it equally. Sumeer comes from a privileged background; he has a rich life, many friends and opportunities, and he is constantly receiving presents from his wealthy relatives. Mary lost her parents when she was young. She lives in a group home, where she has few opportunities, is rarely treated with kindness, and receives no presents at all. Intuitively, in this case, it seems that you should give the treat to Mary rather than Sumeer. Why? Because Mary is worse off than Sumeer. She has had a more difficult life, and she has had fewer goods in her life. This suggests a general principle: When we help others, we should prioritize helping those whose lives are not going as well, those who are worse off.

This principle is important for present purposes because, in many ways, human beings have better lives than animals. We create and join clubs, we travel to other places and compare them to home, we reminisce about our personal history and read about the history of the world, we plan for the future and try to contribute to a better one for others, we recognize the principles of morality and evolution and teach them to the next generation. Animals don't do these things. And what they do often doesn't seem that interesting (although, as we will consider in section 8.3, we are beginning to recognize interesting aspects of animals' lives):

> Our lives are richer and more valuable by virtue of our having goods like these in our lives. To have fewer such goods—or to have less valuable instances of them—is to live a less desirable life, a life at a lower level of well-being.[3]

If animals are worse off than we are, and morally we should prioritize helping those who are worse off, it seems to follow that we should prioritize helping animals rather than human beings. To take just one example, more than eight trillion US dollars (10 percent of global GDP) was spent worldwide in 2018 on protecting and promoting the health of human beings;[4] the amount of money spent on protecting and promoting the health of animals was a tiny fraction of this amount. If there are no degrees of moral status, and human beings have better lives than animals, the principle that we should prioritize helping the worse off seems to suggest that this practice is ethically problematic. It seems to suggest that we should spend the majority of our money, and time, helping animals.[5] If, in contrast, we think that current practice is ethically appropriate, or at least not wildly inappropriate, it suggests that we *are* morally more important than animals.[6] Yes, animals have worse lives than we do, but we are morally more important than they are; hence, it is appropriate to prioritize helping human beings. In this way, the claim that it is often ethically appropriate to prioritize helping human beings seems to suggest that, in fact, there are degrees of moral status.

To assess this possibility, consider the life of a sloth, which seems to contain very few goods. They don't create and join clubs, travel to other places and compare them to home, recognize the principles of morality and evolution, and teach them to their offspring. What do sloths do? They sleep up to twenty hours a day and spend the rest of their time hanging upside down in trees, with occasional trips to the ground to eat, mate, and go to the bathroom. This is, to say the least, not a wonderful life. Does it follow that we should stop helping other humans and spend our time and money helping sloths, with the goal of helping them to lead better lives, with more varied and interesting experiences and projects? If you could spend your time and money helping Mary or the sloth, should you help the sloth? Mary is worse off than Sumeer, but she seems much better off than the average sloth. Granting that, I don't think you should spend all your time helping sloths rather than Mary. To see why not, and to appreciate what the present conclusion implies for the extent to which we should help animals, we need to keep in mind the challenge discussed in section 8.1. We need to be careful not to assume that the implications the principles of morality have for human beings are the same as the implications they have for various animals.

Once we recognize that there are no degrees of moral status, we cannot simply apply our understanding of the principles of morality as they pertain to human beings to animals. We first need to assess whether our understanding of a given principle reflects its implications for human beings in particular or its implications for all individuals who have moral status. In the present case, we are considering the principle that it is better to help individuals whose lives are going less well. To the extent that we focus on human beings, we apply this principle based on what it means for *our* lives to go less well. This is critical because the richness of a life is an important component of a good life for human beings. Forming clubs, reminiscing about the past, working for a better future, recognizing the principles of morality, and teaching them to others are some of the components of a better life for us. But to assess the implications of this principle for animals, we need to be careful not to assume the same account applies to them; we need to be careful not to assume that what constitutes a better life for us is what constitutes a better life for animals. Instead, we first need to consider what constitutes a better life, a flourishing life for the animals in question.

A richer life is a better life for human beings. However, sloths do not have the capacities necessary to do and enjoy many of the things that contribute to a richer life. To get sloths to recognize the principles of morality and teach them to other sloths, we would have to change their brains and their bodies, the way they think, the capacities they have, the things they care about, the things they can understand, and the ways they communicate. The end result of this process would not be sloths who are leading better than average lives. It would be individuals who are no longer sloths, individuals who are not living and, possibly, are not capable of living the life of a sloth. We don't do sloths any favors—we do not help them—by turning them into completely different types of being. Instead, we essentially kill them and create from their remains another type of being. This is a critical insight. It reveals that whether a given way of treating some animals involves helping them, or harming them, depends on the types of animals that they are; it depends on their natures and what counts as a flourishing life for animals with that nature.

The example of sloths reveals that there are two ways to evaluate individuals' lives. We can, to give them names, evaluate the *richness* of an individual's life, and we can evaluate to what extent their life is a *flourishing* one for the individual, for their own sake. The richness of an individual's life, as I am using the term, depends largely on how many valuable and complex activities and experiences it includes. The more, and the more valuable and complex the activities and experiences, the richer the life.

How rich an individual's life is, then, can be determined independently of the type of individual that it is, whether it is a sloth, an elephant, or a human being. How flourishing a life is depends on how the life compares to the best life for that type of individual. How flourishing a life is thus depends on the type of being whose life it is. A flourishing life for sloths is very different from a flourishing life for elephants which is very different from a flourishing life for human beings. The fact that a rich life and a flourishing life are different implies that a flourishing life for some animals can be very different for a rich life. In particular, a flourishing life for some animals might not involve a very rich life at all. That is the lesson of the sloth.

The fact that there are these two ways to assess individuals' lives reveals that we need to consider which assessment is relevant to the principle of prioritizing those who are worse off. The claim that, if there are no degrees of moral status, we should spend most of our resources helping animals assumes that "worse off" refers to the *richness* of individuals' lives. This is a natural assumption to the extent that we are considering human lives: richer lives tend to be more flourishing lives for us. But taking seriously the conclusion that there are no degrees of moral status requires us to consider whether moral principles like this one have the same implications for other types of beings. Maybe a life that is more flourishing for us is not necessarily a life that is more flourishing for them.

To take a very different example, it is bad for us, for our own sakes, to accidentally kill another person (e.g., to kill a child who runs in front of your car). Doing so leads to a less flourishing human life, even when we aren't responsible, and even when we couldn't have avoided it.[7] But it does not follow that it is bad for elephants to accidentally step on and kill small animals. Elephants would not be better off—they would not lead more flourishing lives—if we trained them to avoid stepping on small animals (although the small animals would certainly be better off).

The goal of treating others ethically is not to help them to live the kind of life that is good for us, and it is not to turn them into us. And it's not even to help them live lives that are objectively richer. Instead, the goal, once we accept that animals have equal moral status, is to respect their unique subjective existences and, thereby, help them to live more flourishing lives. And that depends on who they are. Sloths give us reason to help them to live a good slothful life. They don't give us reason to enable them to travel the world, to create clubs, and to learn the principles of evolution and morality.

This is not to say that the life an individual sloth or even most sloths happen to live is the best life for them. It might be that changes, even significant changes to the lives of some animals, would increase the extent to which they lead flourishing lives. Sloths, for example, might be allergic to the bark of trees. And this allergy might result in their suffering. If it does, the fact that sloths have equal moral status to human beings implies that we would have moral reason to try to change this, to develop trees that are better for sloths, or spend some of our health dollars finding ways to eliminate or mitigate the allergy. With that said, a few words of caution.

First, there is no consensus on what constitutes a better or more flourishing life for *human beings*. Developing such an account for animals is likely to be even more challenging. Years ago, on a trip in Tanzania, I saw a pack of African Wild Dogs and started wondering how their lives compared to the life of our pet dog, Katy, who was spoiled and I suspect experienced more pleasure and significantly less pain. But African Wild Dogs live in packs, they cooperate with each other to achieve common goals, and, in so doing, they roam a wide area of incredible beauty. Katy spent a good deal of her time on the sofa. And when she was outside, she was frequently on her own, and on a leash. Were they each living the best lives for them, one for wild dogs and one for domesticated dogs? Was it good for the African Wild Dogs to roam the bush, finding their own food, cooperating with each other to do so? Or would it have been better for them to have a roof over their heads when bad weather came? To be served a dinner that contains no bacteria or viruses and comes in a clean bowl? And to not have to kill others in order to survive? The fact that the answers to these questions are anything but clear illustrates to the challenges that arise in trying to determine what constitutes a flourishing life for different types of animals (in section 8.3 we will consider some possible accounts).

Second, even to the extent that we have a sense for what constitutes a flourishing life for a type of animal, it can be difficult to determine whether a given animal of that type is living such a life. Presumably, some degree of pleasure and contentment is good for sentient animals. But it can be difficult to know to what extent animals are having positive experiences. We can, we think, see whether they are suffering. But looking into your pet's eyes doesn't reveal whether it is bored or it's blissfully communing with nature.

Third, to the extent that helping animals involves interfering in their lives, we might make things worse rather than better. This possibility is only increased by the fact that we do not know what counts as a flourishing life for most animals, and we have limited ability to determine whether a given animal is living that life. In the previous example, it would be difficult to determine to what extent the allergy negatively affects the sloths. And, even if we could figure that out, we would need to consider the possible impact of different ways of addressing it. Would treating the allergy cause side effects that leave the sloths worse off? Would other trees be worse for them?

Fourth, improving the lives of some animals frequently involves interfering in the lives of other animals. Sloths are not the only animals that rely on the trees they live in. Would changes to the bark of the trees, or introducing a different species of tree, benefit sloths to the detriment of the insects that live in the trees? To take a different example, I assume being eaten by African Wild Dogs is terrible for gazelles. Does that give us reason to keep the dogs away for the sake of the gazelles? It depends, in part, on how the gazelles would die instead. Would they starve? Would they develop a form of dementia that leads to dramatic suffering which lasts for years? And what about the dogs? What would they eat instead? Would we need to replace their diet of gazelle meat with protein-rich food we grow in a lab?

The conclusion that there are no degrees of moral status suggests that we should be as concerned with the suffering of animals as the suffering of human beings. But, in deciding when to intervene to minimize their suffering, we need to be careful to avoid making them and/or other animals worse off. There is a more fundamental challenge here as well. To what extent does our intervening in the lives of animals itself conflict with their leading flourishing lives? Put simply: does a flourishing life for animals involve their leading lives that are independent of human involvement or interference?

Every morning, I walk our dogs, Oscar and Maddy, in the woods, where they see and chase deer, while I ponder the ethical dilemmas they thereby raise. I assume the experience of being chased by two large dogs is at least somewhat unpleasant for the deer. Does that make it morally problematic, even though they never catch the deer or even get that close? Alternatively, is being chased and running away part of a characteristic life for deer, suggesting that it might not be so bad or, perhaps, it might be positively good for them, helping them to engage in the "real" life of a deer? One day, they sniffed out a young deer lying on the ground. As they approached, the deer didn't get up and, as I approached, I noticed that it was trying to get up, but couldn't; its back leg was broken and twisted at a sharp angle to the side. I then walked the dogs to the nearby park police station, explained the situation, and suggested they send someone out to check on the deer, with the idea that they might put it down if it was suffering significantly.

After listening to my story, the officer explained that they would not send someone out because the park police try to interfere as little as possible in the lives of the wildlife. Her response made me realize that we both wanted to do the right thing for the deer, but we had very different ideas of what that involves: I wanted to minimize the deer's suffering; the park police wanted to minimize the extent to which humans interfere in their lives. Of course, given the impact that we have had on the environment, it is not possible for deer to live a life entirely free of human influence. The air they breathe, the temperature of their environment, and the number of trees in the woods all have been influenced by us. Still, one might think that less human influence in their lives is better for animals, for their own sakes, in which case my attempts might have resulted in the deer suffering less, but living a less flourishing life.

Briefly summarizing the present section, the conclusion that there are no degrees of moral status does not suggest that we should spend all our time and resources helping animals. The reason is that the principle of helping the worse off refers to individuals who are leading less flourishing lives, not individuals whose lives are less rich. The fact that sloths do not create and join clubs, the fact that they do not understand and teach the principles of morality and evolution, does not give us reason to help them. In contrast, the fact that an animal is suffering does give us reason to help. But as the story of the injured deer illustrates, even here we need to be careful. While the absence of pain and the presence of pleasure are undoubtedly components of a flourishing life for sentient animals, we need to consider whether there are other components and, if so, which ones are more important. To what extent, for example, is it valuable for animals to lead lives independent of us and our lives, even when doing so involves less pleasure and perhaps more pain? We will take up these questions next, in section 8.3.

8.3. Do Rabbits Have Rights?

Sentient animals have moral status; their experiences matter morally. If there were degrees of moral status, the moral importance of animals' experiences would depend on the nature of the experiences factored by their level of moral status. If human beings' moral status were one thousand times greater than the moral status of rabbits, for example, it would be one thousand times worse to torture a human being compared to torturing a rabbit to the same extent. The conclusion that there are no degrees of moral status reveals that individuals' experiences have the same moral importance. Pain and suffering are just as bad when they occur in rabbits compared to when they occur in human beings; pleasure is just as good.

On some moral theories, such as Utilitarianism, pain and pleasure, or positive and aversive experiences more generally, are all that matters morally: positive experiences are good, aversive experiences are bad. In fact, there is more to morality than simply maximizing the amount of pleasure or positive experiences in the world; there is also respect for individuals. Some people think that respect for individuals is exhausted by not interfering in their lives, that individuals have only negative rights against being treated badly, against being subjected, for example, to pain and suffering. Others think individuals also possess positive rights which give them claims to the help and assistance of others. For present purposes, we do not need to try to settle this debate. Instead, the important point is this: the conclusion that there are no degrees of moral status implies that the experiences of animals matter as much morally as the experiences of human beings. Hence, to the extent that human beings have rights which protect their experiential interests, animals have the same rights. Humans have a right against being tortured: torturing human beings is either always wrong, or it is permissible only in the extremely unlikely event that torturing some humans is the only way to avert a horrific outcome, such as the deaths of millions of people. The conclusion that there are no degrees of moral status, together with the fact that dramatic suffering is also very bad for animals, reveals that animals have this right as well (in section 8.5 we will consider the implications of this conclusion for animal experimentation).

In addition to rights which protect our experiential interests, human beings also possess rights based in appropriate respect. We have strong rights to direct the course of our lives: where we live, whether and with whom we mate, whether we have children. Of course, respecting human beings in these ways often helps to promote our experiential interests. We get frustrated and angry when others try to control our lives. Furthermore, we often know what we enjoy and what we don't. Our right to direct the course of our lives is thus important for promoting our welfare. But our right to direct the course of our lives is not simply a means for improving our experiences; it is a matter of ensuring appropriate respect for us. Even when we make decisions which lead to our having negative experiences, even when we make our lives worse, we still have a right to decide for ourselves.

Given that animals have the same moral status as we do, it might seem to follow that animals have the same strong rights: rights to direct the course of their lives, where they live, whether and with whom they mate, whether they have children. And that might be just the beginning. We have a strong right against being killed.[8] We have a right to decide whether and when we take walks, where and what to eat; others are not allowed to own us. From the conclusion that there are no degrees of moral status, it seems to follow that dogs and deer, pigs and chickens, alligators and elephants similarly have a right to live where they want, to move around as they want, to mate with whom they choose, to not be owned by us, and to not be killed for food. We cannot neuter them, displace them with oil fields or roads, build our homes on top of theirs. If that's right, we would have to give animals free rein and learn to live with the results. Moreover, to the extent that we also have positive rights, animals would have them too. For example, the fact that we have rights to be part of the political community, to share in its benefits, suggests that animals should be treated as equal citizens.[9] Similarly, some have concluded that in a world without degrees of moral status: "It is wrong to conduct any experiment on a sentient non-human which we would regard as immoral were it to be conducted on a human."[10]

Of course, animals do not have exactly the same respect-based rights as humans. Humans have a right to a trial by a jury of our peers. Presumably, rabbits do not have that right, even though they have the same level of moral status. With this in mind, proponents of animal rights frequently argue that animals have the same *fundamental* rights as we do, where fundamental rights provide protection for the most important or central aspects of our lives. This includes, but goes beyond, rights that protect our experiential interests. It also includes respect-based rights: a right to life, liberty, procreation, and freedom to direct the course of one's own life.

The conclusion that there are no degrees of moral status would imply that animals have the same fundamental rights as human beings if it were the case that fundamental rights "co-travel" with moral status in the sense that individuals who have the same level of moral status necessarily have the same fundamental rights. However, to the extent that individuals' moral status is a result of their being sentient, there is no reason to think that fundamental rights will co-travel with moral status in this way. Sentience is important morally because it implies that individuals have a stake in how they are treated and how their lives go; it implies that there are better and worse ways for their lives to go for their own sakes. But the fact that individuals are sentient does not determine *how* they ought to be treated; it does not determine which types of treatment and lives are better and worse for them. Respect for the subjective existences of sentient animals does not determine, for example, whether they have a strong right to determine with whom they mate, or whether they have children. And, to repeat a lesson learned earlier: we should not assume that what is valuable for us is necessarily what is valuable for them. The fact that it is important for us to determine whether we have children does not necessarily imply that it is valuable for rabbits.

Having moral status implies that how animals are treated matters morally, that there are moral reasons to respect them and to allow them to live a flourishing life.[11] But the fact that an individual is sentient tells us very little about what constitutes a flourishing life for them, except that suffering is bad for them, for their own sake, and pleasurable experiences are good. In a world without degrees of moral status, then, the only rights that co-travel with moral status are rights against being subjected to significant aversive experiences and, if one endorses positive rights, rights to be helped to at least a basic level of positive experiences.

To determine whether animals also have a right to direct the course of their lives, some commentators identify which capacities enable us to do so and then assess whether animals possess these or similar capacities. For example, Tom Regan, a proponent of strong animal rights, points out that all sentient animals are the subjects of a life and, therefore, deserve respect. He recognizes that not all sentient animals have the capacities necessary to direct the course of their own lives; for example, some do not have a sense for the future or a sense for their own future. But many animals have the necessary capacities; they have what Regan calls a "psychophysical identity" over time, along with the ability to initiate actions in pursuit of their desires and goals. These animals, he argues, have a strong right to direct the course of their lives.[12]

The capacities necessary for directing the course of one's own life come in degrees. Some have a rich sense for their future; others have a dim sense of it. The strong rights view that Regan endorses thus requires some threshold on how much of the underlying capacities individuals must possess in order to have a strong claim to direct the course of their lives. The challenges that we considered in section 3.5 for identifying morally significant thresholds on the possession of superior cognitive capacities thus undermine this view as well. Alternatively, proponents might argue that simply possessing the necessary capacities to *any* extent endows individuals with a strong right to direct the course of their lives. This seems implausible though; it seems implausible to think that an individual who has the most rudimentary capacity to direct the course of their own lives has a right to do so that is as strong as the analogous right of competent adults.

More importantly, the reason why we have a strong right to direct the course of our lives is not simply because we possess the relevant capacities to do so. A flourishing life does not involve the exercise of all the capacities one happens to possess. Human beings have the capacity to juggle, but we can live fully flourishing lives without ever exercising this capacity. Why is that so obvious? Because we live our lives from the inside and that gives us insight into what is important for us and our lives. We also communicate with other human beings, we talk to our contemporaries, and we read about our predecessors, to see what they thought and the extent to which they agree. Over time, we develop a shared sense for at least some aspects of a flourishing life for human beings, including the importance of juggling and the importance of directing the course of our own lives.

To determine whether sentient animals have a strong right to direct the course of their lives, then, it is not enough to determine whether they have the capacity to do so. We need to identify the most important or central aspects of a flourishing life for them and consider whether this is one of them. Of course, animals are very different from humans, and animals are very different from one another. The assumption that at least the most important components of a flourishing life for human beings are likewise important components of a flourishing life for animals thus runs the risk of failing to respect animals as distinct individuals. It fails to take seriously their separateness and distinctiveness. Doing so also raises the possibility that some things which are critical to flourishing for animals are not important for us, raising the possibility that they have rights we don't possess. A right to a trial by jury is important for us, but not them. There might similarly be things which are important for them, but not us.

The magnitude of the challenge involved in determining what constitutes a flourishing life for animals is illustrated by the debate over the central aspects of a flourishing life for human beings. Some people believe that human flourishing involves maximally making our own decisions, whatever they might be, without the influence of others. Others believe that an independent life is a barren one, that flourishing for human beings is possible only when our lives are deeply enmeshed in the lives of others, our spouses, families, friends, and neighbors, to the extent that they make decisions about important aspects of our lives for us. Determining precisely what constitutes a flourishing life for animals is likely to be even more difficult.

One way to define a flourishing life for different animals is in terms of *species typical functioning*. The basic idea here is that the best life for animals is the life that a given species tends to live in the wild, the life that is characteristic for that species. An important virtue of this account is that it recognizes the difference between a rich life and a flourishing life noted in section 8.2. Sloths, as we saw, spend most of their time sleeping, and the rest of it hanging upside down in trees, with occasional trips to the ground. This represents a fairly impoverished life. But on the present account, this is nonetheless a flourishing life for sloths because it is characteristic of sloths. This approach thus has the significant virtue of defining what constitutes a flourishing life in a way that respects the differences between animals and us, and their differences from each other. The best life for sloths, rabbits, and dolphins is the life that members of each species tend to live.

The lives animals tend to live have been significantly shaped by evolution, which results in animals evolving in ways that lead to the proliferation of their genes and the survival of their species. Depending on the circumstances, especially the local environment and the competition a species faces for space and resources, what leads to success evolutionarily might be consistent with a flourishing life for the individuals within the species. For example, a species can survive only if (a sufficient number of) its members have food and live to the point where they reproduce. Having enough food likewise tends to be good for individuals, for their own sakes.

In other cases, what leads to success evolutionarily might not be consistent with a flourishing life for the individuals within the species. For example, the lives of many species are defined by hierarchical social arrangements, with some individuals having dominant roles over others. Wolves live in packs and the alpha male and female are the only ones permitted to breed, and the best food is reserved for them. This arrangement might have helped the species to survive. But the fact that the likelihood of the species surviving increases when most of its members do not breed, and eat leftovers, does not imply that living that life is better for the individual members. To take a more extreme example, the lives of some animals are characterized by violent and forced sexual intercourse. From an evolutionary point of view, this, too, could make sense. It might have led to there being more members of the species, thus increasing the chances that the species survives.[13] But the fact that this contributes to the survival of the species provides little, if any, reason to think it is good for individual members to be subjected to forced and violent intercourse.

Humans also frequently treat each other badly: we yell, we argue, we fight, we kill each other. These are, unfortunately, characteristic behaviors. But they are not part of a flourishing life, much less central to a flourishing life, for us. Instead, we recognize that these behaviors get in the way of our living flourishing lives: the fact that they are characteristic provides all the more reason to try to eliminate them, not reason to preserve and respect them. Perhaps just as importantly, what constitutes a flourishing life for human beings is not necessarily limited to the capacities with which evolution has endowed us. As we will consider in section 8.7, many people think that altering our capacities using genetic technologies could dramatically increase human flourishing. For example, no human being has the capacity to live for 250 years, or forever. But some people think we would be much better off if we did. Or if we could remember everything we have ever learned (without getting confused or muddled). Or if we could selectively eliminate memories which we find unpleasant rather than allowing them to haunt us for years. These possibilities underscore the difference between a characteristic life and a flourishing life.

The philosopher Martha Nussbaum suggests that, in order to figure out what constitutes a flourishing life for a given animal species, we will need to live with them for years, with love and sensitivity, learning from them as they "express their deepest concerns while they try to live."[14] While this sounds right, it also underscores the challenges we face here. To determine what constitutes a characteristic life for animals, Nussbaum recommends that we look to the things they strive to do. But gleaning what animals are striving to do from how they act isn't easy.

First, we have to figure out whether they are in fact striving do to anything at all, as opposed to simply acting out of instinct or acting without really thinking about it or caring. Second, to the extent they are striving to do something, what is it? This is difficult because any given set of actions can represent an attempt to do many different things. For example, animals frequently engage in the actions associated with mating and reproducing. They might thereby be trying to have offspring. Or simply to have sex. Or to do something that feels good. Or to interact with others. Or to express caring. Or to express power and domination. Third, even if we can figure out what they are striving to do, we still need to figure out whether doing those things is part of a flourishing life for them.

The fact that animals strive to do a certain thing, even repeatedly, does not establish that doing that thing is part of a flourishing life for them. Humans, as we noted previously, repeatedly attempt to do many things that are inconsistent with rather than part of a flourishing life, such as seeking revenge, being cruel, or starting wars.

A different way to understand the moral significance of a characteristic life is in terms of its protecting animals from our interference and depredations. Consider a species that develops new behaviors in response to the trees in its environment getting taller and casting longer shadows and more shade on the plants they eat. Or consider an animal species that alters its behavior in response to another animal species moving into its habitat. While these changes might lead to significantly different lives for the animals, they do not seem to lead to less characteristic ones. Instead, these changes seem like characteristic responses to changes in the animals' environment.

Now consider changes that occur in response to *humans* moving into the animals' environment. Instead of their grasslands becoming shaded by maturing trees, they are shaded by a strip mall. Instead of a different animal species moving in and setting up home, it's a new housing development. The animals' responses to these changes, even if they are similar to the changes made in response to a different animal species moving in, seem more like responses to interference with their characteristic lives as opposed to characteristic responses to changes in their environment. This understanding is consistent with the common view that interactions between different animals and the environment are part of their natural lives, whereas interactions with human beings represent disruptions to their natural lives.

Speaking of the wild Osprey that live in their parks, rangers in the state of Maryland "do not interfere with their natural cycle which sometimes might include interactions with predators or natural disasters"[15] This view regards what Mother Nature does to animals, and what animals do to each other, including attacking and eating each other, as part of a natural or a characteristic life. In contrast, the rangers get involved when humans attempt to influence the lives of the Osprey, including attacking and eating them. Those interactions are regarded as inconsistent with a natural or characteristic life for the Osprey.

This points to a different way to understand the value involved in animals living a characteristic life. A characteristic life, in the sense of the life that a species leads in the wild independent of human influence, might not always be best for its individual members. But living that life precludes our interference. Consider capturing an animal in the wild, where it is threatened by predators and does not always find enough to eat, and maintaining it in pleasant captivity, where it has sufficient room to move around, good food and water, a comfortable environment, and conspecifics to live and play with. Is this problematic? Is it bad for the animals themselves, for their own sake, to have the course of their lives so fundamentally shaped by humans? Or does it depend on whether the animals have more pleasant experiences than they would have had in the wild?

We have been discussing these questions in terms of what is better and worse for the animals, for their own sakes. However, to the extent that these questions involve how we interact with animals, it is important to keep in mind the possibility that our responses might be influenced by what we prefer for ourselves. Most importantly, in addition to avoiding bad outcomes themselves, human beings prefer to avoid being responsible for bad outcomes. When given the option between trying to help others or staying out of their lives, we frequently choose the latter, not because it is better for them, but because it feels better for us, it protects us from feeling responsible if things don't turn out well. This raises the following question: To what extent is the claim that animals are better off leading characteristic lives which are independent of human interference informed by what is best for the animals for their own sakes? And to what extent is it informed by our desire to avoid feeling responsible for how their lives go?

One final option. Earlier we noted that individuals can possess different types of value. Victoria Falls has value as a wonder of nature, and *Mona Lisa* has aesthetic and historic value. The value of *Mona Lisa* traces, in part, to the fact that it was created by human beings (its value would be different if it had been created by an artificial intelligence program), whereas the value of Victoria Falls as a wonder of nature traces to the fact that it wasn't. Animals also represent wonders of nature, amazing and wonderful adaptations to the natural world. This raises the possibility that the problem with human interference in animal lives is that it undermines the extent to which they are wonders of nature. This suggests that, in some cases, there might be a tension between doing what is best for animals for the own sakes versus maintaining their value as part of the natural world.

Some animals eat others to survive, which leads to dramatic suffering for the individuals that are eaten. In theory at least, we might be able to reduce or even eliminate this suffering by eliminating predation through selective breeding, genetic manipulation, or other methods. That might dramatically improve the lives of sentient prey. It might also not harm the predators if we can identify other ways for them to satisfy their nutritional needs. For the sake of the animals themselves, then, the elimination of predation could be a very good thing.[16] But it would also dramatically reduce the extent to which the affected predators are products of and part of the "natural world," independent of the interference of human beings. In order to determine how we ought to treat animals, then, we need to keep in mind that what is good for them, for their own sakes, may not always be what promotes their value in other ways.

The present discussion highlights the challenges we face in determining what constitutes a flourishing life for different animal species. The fact that this is so hard reveals that it will be extremely challenging to determine what is involved in respecting them. Granting that, one might argue that animals can flourish only if they are alive. This, combined with the fact that sentient animals have the same moral status as we do, seems to suggest that, in addition to a right against being caused to suffer, animals also have a right to life, much like human beings.

Death is bad for us because it involves losing all that we have and being deprived of all that we would have had. Death has the same impact on rabbits. Hence, even though rabbits' lives contain less of value than our lives, the death of a rabbit seems just as bad for the rabbit as our deaths are bad for us:

> Even if the rabbit's life is not as important to her as yours is to you, nevertheless, for her it contains absolutely everything of value, all that can ever be good or bad for her, except possibly the lives of her offspring. The end of her life is the end of all value and goodness for her.[17]

Similarly:

> So it is, one might say, with death, which forecloses all further experiences and thwarts all opportunities for satisfaction. If A and B both value living and are in the prime of life, they would be harmed equally by death, no matter how "dull" B is in comparison to A.[18]

These passages suggest that rabbits have a right to life equal to ours. After all, death results in the same loss for them, the loss of essentially everything, as it does for us. This seems right to the extent that we consider lives from the inside. For the rabbit, from their perspective, their life contains essentially all that is of value for them. And from inside my life, it contains much of what is of value for me. These subjective experiences, these unique existences, deserve respect. The value from inside the life is necessarily connected to the individual themselves, which seems to preclude comparisons of the value of the two lives from some neutral perspective. Without appeal to such a perspective, it does seem that we have no basis to claim, as I have, that there is more at stake in our lives and our deaths than those of rabbits.

The possibility that we can evaluate the value of different lives is suggested by the fact that the lives of individuals do not necessarily contain all that is of value for them. For humans, for example, the lives of their children, whether their children flourish or not, can be of great importance for the parents. This reveals that we evaluate lives based on factors that are not contained in the lives themselves. And that points to the possibility of judging the value of individual lives from outside those lives.

To step away from animals for a moment, the life of a given tree contains all that is of value for the tree, and killing the tree eliminates all that value. Still, it is worse to kill human beings than it is to kill trees. Of course, trees are not sentient; they have no awareness of the value that is contained in their lives and, hence, there is no sense in which killing them is bad for the tree for its own sake. But notice that this already involves judging the comparative value of different lives from outside the lives in question. Specifically, sentient existence is more valuable for the sentient individual than nonsentient existence is for that individual. Put another way, more is lost when sentient individuals die than when nonsentient individuals die.

Now consider an individual that perceives its environment, but does not have any subjective experiences. This might be what life is like for some insects.[19] They are aware of the environment in ways that allow them to navigate and avoid running into things. However, running into things does not produce an aversive experience (although it might result in physical injuries). Here, too, the experiences contained in the life of the insect are all the insect has. And the insect is aware of the world. Nonetheless, comparing the richness of our lives to theirs reveals that killing these insects is not as bad as killing human beings because it leads to much less loss.

Similar considerations apply within humans. Imagine a hospital that has two patients who are unable to breathe on their own, but only one ventilator. Who should be offered the ventilator? Both patients are thirty-five years old and happy. One patient, if she is put on the ventilator for a few days, is expected to live another fifty years. The other patient has untreatable cancer which will end his life in the next few months, whether he is put on the ventilator or not. The lives of these two individuals, and the time they have left, is all that each one has. Does it follow that it is impossible to make comparisons between the two lives and determine who ought to get access to the ventilator? No. Imagine the second patient says that the ventilator should be given to the first patient because she has longer to live; hence, the rest of her life has more value for her than the rest of his life has for him. Even though both will lose their lives and all they have left, the first patient will lose more. This does not only make sense, it seems right. And that suggests that, although these comparisons are anything but easy, and likely impossible in some cases, they are possible in others.

Briefly summarizing this section, the fact that we have certain rights, and rabbits have the same moral status as we do, does not imply that rabbits necessarily possess the same rights as we do. To make that determination, we need to know two things: what respect requires when it comes to rabbits; what constitutes a flourishing life for rabbits and how important is that life, for rabbits, for their own sake. While significantly more work will need to be done, then, the answer to the title question of this section is *Yes*, rabbits do have rights, but they do not possess exactly the same rights we possess. We have seen that rabbits have a right against being subjected to significantly aversive experiences that is equal to ours, but they either do not have a right to life or their right to life is significantly weaker than ours.

8.4. The Ethics of Raising Animals and Eating Them

Approximately eighty billion animals are killed every year so that humans can eat them.[20] The present section discusses the ethics of this practice in light of the conclusion that there are no degrees of moral status. Does the fact that rabbits have the same moral status as we do imply that this practice is unethical? Alternatively, does the fact that rabbits do not have a (strong) right to life imply that it is acceptable?

Before we try to answer those questions, it is important to note that the effects of raising and eating animals go beyond the impact on the animals themselves.[21] Raising animals for human consumption uses significant amounts of water and releases toxins into the air. Eating animal meat, at least in significant quantities, might be bad for our physical health, and killing animals might be bad for workers' mental health. While these issues are important, I will set them to the side. The question for the present section is the following: If we raised animals that are not unhealthy for us to eat, and we learned to do it in ways that do not have a negative impact on the environment, the individuals who work in the meat industry, or others, would it be ethically acceptable? Or is the practice itself ethically problematic? Answering this question is vital to assessing current efforts to develop meat alternatives. How important these efforts are, and how much we should be willing to pay for them, depends critically on the ethics of raising animals and eating them.

As it is currently practiced, creating animals, raising them, and then eating them involves subjecting billions of animals to significant suffering: raising them in confined spaces, mistreating them while alive, and killing them in ways that cause significant pain. All of this suffering yields two primary benefits. First, there are the sensory benefits that many people derive from eating meat, although the development of meatless substitutes, such as vegetarian burgers which taste like meat, is reducing, and might soon eliminate the extent to which meat is necessary to realize these benefits. Second, eating meat is central to maintaining and partaking in a range of social, cultural, and religious practices and traditions.

These benefits, in terms of sheer magnitude, are dramatically less than the harms caused by current methods of factory farming.

Many people, I suspect, agree, but continue to support factory farming, or at least do not object to it, because they believe that we have higher moral status than animals. If there were degrees of moral status, this view might turn out to be right. The harms of current practice are many orders of magnitude greater than the benefits. But so, too, might be the difference between our moral status and the moral status of animals. At a minimum, without a reliable way to measure levels of moral status, this view is not obviously mistaken. The present conclusion that there are no degrees of moral status changes this calculation completely. We do not multiply the benefits for human beings by the extent to which our moral status exceeds the moral status of the affected animals. We simply compare the benefits for us to the harms for them. The fact that the harms dramatically outweigh the benefits thus reveals that current approaches to factory farming are unethical. That is the first conclusion of the present section. It gains support from the fact that respect for individuals as individuals applies to animals. While the implications of proper respect for animals are not entirely clear, essentially torturing individuals for the opportunity to have better tasting food, and to maintain some of our traditions, is not consistent with it.

This brings us to humane animal farming. Imagine a farm that breeds rabbits, treats them well during their lives, and then kills them painlessly. Undoubtedly, rabbit meat produced in this way would be more expensive, but enough people might be willing to pay for it to make this type of animal farming sustainable. Assuming it is, what does the fact that there are no degrees of moral status imply about whether it would be ethically acceptable?

Even humane animal farming involves killing animals, which deprives them of the future experiences they otherwise would have had. If those experiences would have been pleasurable or otherwise good for the animal, killing them deprives the animals of those goods. Doesn't it follow that killing animals for food, even when they are raised humanely, is ethically problematic?[22] As noted in section 4.2, individual lives involve essentially two things: the impact the world, including others, has on the individual, and the impact the individual has on the world, including others. The impact the world has on individuals involves their experiences. All sentient individuals have experiential interests: they have an interest in having positive experiences, and they have an interest in avoiding aversive ones. The impact we have on the world and others involves our contributions. Our lives go better for us for our own sakes when we are involved in and contribute to valuable projects, and our lives go worse for us when we are involved in and contribute to detrimental projects. This is because we have contribution, as well as experiential, interests.

As we saw in section 4.2, experiential interests are contingent in the sense that promoting them is good for an individual to the extent the individual is alive. This is perhaps most clear with respect to individuals who truly live in the moment. They experience whatever is happening to them at the specific moment, and then experience whatever happens to them next, without any connection between the two moments. These individuals have no plans for the future, no projects or activities that extend over time. As a result, they have experiential interests, but no contribution interests, in which case killing them painlessly does not harm them at all.[23]

The fact that individuals are not harmed by being deprived of future positive experiences they otherwise would have enjoyed does not imply that killing them painlessly is necessarily acceptable. The reason is that killing individuals, even painlessly, can undermine their contribution interests. Killing someone who is in the middle of making a beautiful piece of pottery eliminates the opportunity to complete the project and, thereby, harms them. Killing human beings also eliminates the possibility of their contributing to valuable projects in the future. It seems reasonable, in contrast, to assume that rabbits have minimal, if any, contribution interests. The same is likely true for at least a number of animals that are raised for food, including chickens and fish. If that is right, humane farming of these animals does not harm them. Animals with greater cognitive capacities, including pigs and octopuses, might well have contribution interests, in which case, raising and eating them, even humanely, would harm them. These animals might also have at least some claim to direct the course of their lives, which is contradicted by killing them, even painlessly.

Breeding and raising animals involves entering into relationships with them, which have their own norms on appropriate treatment. Consider our relationships to our pets. We don't eat our pets even after they die a natural death. Why not? Eating our pets after they die does not harm them and does not violate respect for their agency. The reason why it is nonetheless problematic to eat them is that being in a relationship with a pet is inconsistent with regarding them as a source of food, even after they die. To fully assess the ethics of humane animal farming, then, we need to determine what constitutes appropriate relationships with the animals, ones that are not our intimates but not wild either. Unfortunately, for at least two reasons, I do not think that we are not currently in a position to develop such an account.

First, the norms that govern relationships are not things that one can see simply by looking at the individuals involved in the relationship, nor things that one can discover by pure philosophical reflection. Neither of those approaches would reveal that it is inappropriate to eat our pets after they die. We figured that out by living in the relationships, by having pets and learning what constitutes appropriate treatment of them. In contrast, what at least most of us regard as appropriate relationships with other animals have been strongly influenced by our history, and by the belief that they have lower moral status than we do. This undermines our current ability to identify the appropriate norms for our relationships with animals.

This challenge is reflected in our habit of appealing to the norms of human relationships to determine how we ought to treat animals. Those are the only models we have for what qualifies as appropriate relationships with moral equals. To do better, we will have to live with animals with the understanding that they have the same moral status as we do. Humane animal farming illustrates the challenges here. Consider a gourmet cook who breeds chickens, gives them a good life, and then kills them painlessly, after which she makes delicious roast chicken dinners for her friends and neighbors. This practice does not harm the chickens and involves, at most, minimal violation of appropriate respect for them as individuals and for their agency. To this extent, the pleasure that the dinners engender, along with the opportunity for the chef to develop her skills and benefit her friends, seems to justify this practice. But we also need to consider whether this involves an appropriate relationship between the woman and her chickens. Is it appropriate for her to treat moral equals that she raises in this way? Is it appropriate for her friends to enjoy the meal, or should they be troubled by the fact that it required the killing of living beings that were enjoying their lives? Alternatively, should they be glad that breeding and raising the chickens for dinners provided them with a good life for some time? One way to test whether humane animal farmers consider their animals to be moral equals is to assess how they regard the animals' experiential interests. Put simply: can we raise animals and treat their experiential interests as equal to ours, but then kill and enjoy eating them?

Finally, a full analysis of humane animal farming would need to take up the issue I set aside at the beginning of this section: the impact breeding, raising, killing, and eating animals which share our level of moral status has on us. Even when it causes the animals no harm, is consistent with appropriate respect, and is consistent with having an appropriate relationship with them, this practice involves our destroying, on a routine basis, complex individuals with inherent value. To what extent is that, and to what extent should that be problematic for us? For example, to what extent does it cause the animal executioners psychological harm? To some extent that depends on the norms of appropriate relationships with them. So until we make that determination we might not be able to answer the present question. Some types of understanding come not from thinking and analysis, but from living with awareness.

In summary, we have been considering the implications of the conclusion that there are no degrees of moral status for the ethics of raising animals and eating them. The fact that animals have the same level of moral status implies that their experiential interests count as much as ours. This suggests that at least most methods of raising animals for food, which cause significant suffering without anything close to corresponding benefits, are unethical. We then considered the ethics of humanely raising animals for food. If the animals are living a good life, killing them eliminates their future positive experiences. However, those are contingent interests that benefit animals to the extent they are alive. Elimination of those experiences does not provide a reason why it is wrong to kill animals.

We then considered whether raising animals and killing them painlessly for food is consistent with appropriate respect for them. The question here is whether it is better, more respectful to:

a. Bring some rabbits into the world, raise them to have good lives, kill them painlessly, and eat them.
b. Not bring the rabbits into the world in the first place.

Granting that animals deserve respect as individuals, we have only just begun to figure out what that involves. As we do, it will be important not to assume, as noted previously, that what constitutes appropriate treatment of human beings constitutes appropriate treatment of animals as well. It is inappropriate to raise some human beings and kill them for food, even if the alternative would have been their not existing at all. Is it similarly worse to breed rabbits and kill them for food rather than not bringing them into existence in the first place? And, of course, appropriate treatment of animals is likely to vary widely depending on the animals in question. The capacities of chickens give them at best a minor claim to respect for their agency, to allow them to direct the course of their own lives. In contrast, the cognitive capacities of other animals, including pigs and octopuses, are much more sophisticated in this sense, suggesting that they have a moderate claim to direct the course of their own lives, which is inconsistent with killing them for food.

Finally, we need to be aware of the impact the killing has on us. I do not believe that painlessly killing chickens harms them to any significant extent. At the same time, I would personally find it difficult to kill a chicken. Is that inconsistent? Does my concern suggest that I really think killing chickens is morally problematic? Or does it trace to the fact that I am mistakenly thinking of killing chickens as akin to killing human beings? The answers to those questions will require living with animals in recognition of the fact that they have the same level of moral status as we do but are also different from us in many ways.

8.5. The Ethics of Animal Experimentation

Regulations governing research with animals are very different from regulations governing research with humans. These differences are especially stark when it comes to studies that pose risks but do not offer subjects the potential for personal benefit, so-called *net-risk* studies. An example is taking a skin biopsy from healthy subjects to study in the laboratory, exposing the subjects to some risks without any chance for them to benefit. Every year, in the United States, hundreds of thousands of animals are enrolled in net-risk studies in order to collect data that might benefit humans. Of these, a significant minority involve subjecting the animals to pain without relief. The pain the animals experience ranges from mild to intense, and lasts minutes to days.

For the past fifty years, these studies have been governed by regulations modeled on the "three Rs" which were first described by Russell and Burch in 1959.[24] The three principles, replacement, reduction, and refinement, direct researchers to design their experiments in ways that cause as little harm to the animals as possible. Replacement directs researchers to use non-animal methods, for example, experiments in test tubes or computers, whenever possible. Reduction directs researchers to use the fewest number of animals needed, while refinement mandates the use of methods that produce the least amount of pain, suffering, and distress, consistent with obtaining the data needed for the study.

Net-risk studies with human subjects are governed by similar principles. Investigators are directed to enroll the minimum number of subjects, and to design the experiments to minimize the risks. However, unlike research with animals, there are limits on the risks and harms to which human subjects, especially those who cannot consent, may be exposed for the benefit of others. To see how this difference plays out in practice, consider research on septic shock, a condition which involves widespread infection, frequently leading to life-threatening complications.

Little is known about how to treat septic shock, and the mortality rate is high; approximately 40 percent of adults who get septic shock die of it. To test potential new treatments for septic shock, researchers first developed an animal model for the condition, in dogs.[25] Developing an animal model involved giving septic shock to the dogs and then testing possible treatments on them. The symptoms of septic shock are uncomfortable and, at times, intensely painful. In addition, much of what is medically important about septic shock—heart rate, blood pressure, level of pain—is affected by analgesics. Thus, to yield scientifically valid results, these experiments are conducted in dogs that are kept in pain for hours, even days. This level of distress and pain would never be approved in research with competent adults, much less in humans who cannot give informed consent. The reason why this research was approved in animals, but would never be approved in humans, is that existing regulations assume human beings have significantly higher moral status than animals; hence, we deserve significantly greater protections.

The conclusion that there are no degrees of moral status might seem to have obvious implications for net-risk research with animals.[26] If animals have the same moral status as human beings, and animals cannot consent, they should not be exposed to *any* risks or burdens for the benefit of others. In other words, we should abolish net-risk research with animals. While this view makes sense, and it has been endorsed by some commentators, it does not involve providing animals with the same protections as humans; it involves providing animals with greater protections. Regulations and guidelines around the world permit human beings to be exposed to some risks, even without their consent, provided the research is valuable and the risks are not excessive. Hence, the fact that animals have the same moral status as humans does not imply that it is unethical to expose them to any research risks for the benefit of others. It implies that it is unethical to expose them to excessive risks. It thus poses the challenge of determining what level of net risks is acceptable and when they are justified by the value of the study.

Currently, review committees are not required to make these assessments. They do not have to assess how much the experiment might frighten the animals. Or cause them anxiety. Or pain. They also don't have to estimate the chances that the experiment will ultimately benefit human beings and whether this chance justifies the harms to the animal subjects. Because humans are assumed to be significantly more important than animals, any chance to benefit us is regarded as sufficient to justify essentially any level of suffering to them. The conclusion that animals' moral status is equal to ours implies that this approach is ethically problematic. To determine whether net-risk studies with animals are ethically appropriate, review committees need to assess whether the potential benefits justify the risks. To do that, they need to estimate the extent of the risks and harms to the animal subjects, the extent of the potential benefits to human beings, and compare the two.

I have argued that death itself is not a significant harm for many animals. This suggests that these assessments should focus primarily on protecting animal subjects from pain, suffering, and other aversive experiences, as opposed to trying to avoid death and keeping them alive as long as possible. Assessments of the extent to which a given experiment is likely to cause animal subjects to suffer should take into account the fact, noted in sections 5.1 and 5.2, that human beings have a tendency to downplay the suffering of animals. Simply recognizing this possibility might help to a certain extent. Beyond that, it will be important to develop better measures for the suffering of animals and to consult experts who are familiar with the animals in question, people who know how the animals typically behave and what different behaviors suggest regarding what they might be experiencing and how they are feeling.

Like research with animals, pediatric research involves exposing individuals who cannot consent to some risks in order to learn things that might benefit others. Regulations on research with children might thus provide a model for identifying what level of net risks is acceptable in research with animals. Unfortunately, there is no consensus on how much net risk is acceptable in pediatric research. Most importantly, prominent moral theories clearly get it wrong. Consequentialist theories, which focus on maximizing overall benefits, suggest that there should be no upper limits on the magnitude of the harms to which pediatric subjects can be exposed. Instead, any level of risks and harm can be acceptable provided the potential benefits to others justify them.

This approach has the potential to expose children to unacceptably high risks. It could, for example, justify enrolling children in the septic shock experiment and exposing them to excruciating pain for weeks if doing so offered sufficient potential benefit for others. No regulations allow that, and none should allow it. In contrast, moral theories which emphasize the importance of respect for individuals imply that it is unethical to expose individuals who cannot consent to any risks for the benefit of others. These theories suggest that children should not be exposed to any net risks, no matter how low the risks and how important the potential benefits for others. This, too, is widely recognized as a mistake.

Given that leading moral theories do not provide clear guidance, how do we regulate pediatric research? Regulations and guidelines around the world take a pragmatic approach between these two extremes, permitting children to be exposed to some risks, provided the net risks are sufficiently low and they are justified by the potential for the research to benefit others. Regulations in the United States, for example, permit children to be enrolled in net-risk studies that pose *minimal* risks or a *minor increase over minimal* risk.[27] Minimal risks are defined as risks which do not exceed the risks "ordinarily encountered in daily life," while a minor increase over minimal risk is not defined.

US regulations include an additional category which does not place an upper limit on risks to children.[28] Instead, it permits pediatric studies provided they have sufficient social value and they are conducted in accordance with "sound ethical principles." What this permits in terms of the risks to which pediatric subjects are exposed is left to the judgment of reviewers. The important point for present purposes is that these regulations do not specify what level of net risks is, and what level of net risks is not consistent with sound ethical principles. We are thus back where we started. Our goal was to identify an upper limit on risks, pain, and suffering in net-risk animal research. Given that animals cannot consent, we looked for guidance to existing regulations for research with children. Yet US regulations at least do not specify an upper limit.

We could expand our search to consider guidelines and regulations from other countries. However, there is no consensus there either. They span the range of prohibiting net-risk pediatric research, permitting minimal risks, permitting a minor increase over minimal risk, and following the US regulations.

Given that we can't be certain where the limits on appropriate risks are, one approach would be to prohibit the studies which are clearly and seriously problematic. In particular, causing individuals who cannot consent to experience prolonged and significant suffering for the benefit of others is clearly inconsistent with respect for them as individuals. This suggests that regulations for research with animals might be revised to prohibit net-risk studies that pose greater than moderate pain or suffering. To implement this approach, regulations could define three levels of risk: minimal, moderate, and high. Minimal risk studies with animals would be permitted as long as there is reason to think that the potential benefits justify the risks. Moderate risk studies would be permitted when there is *compelling* reason to believe that the risks are justified by the social value of the study.

High-risk studies could be prohibited, or they could be subject to additional procedural requirements similar to the US pediatric regulations which mandate consultation with a panel of experts, opportunity for public review and comment, and approval by a government official. Adopting this approach would make regulations for research with animals similar to existing regulations for research with children. While that fact does not prove this approach is right for animals, it reduces the chances that any bias in favor of humans will lead to significant harm to animal subjects.

Revising regulations for research with animals in these two ways—mandating that the net risks are not excessive and they are justified by the value of the study—will help to make them consistent with the conclusion that there are no degrees of moral status. This conclusion also suggests that the regulations will need to be revised to satisfy the requirement of *replacement*, which mandates that there are no alternative approaches to conducting the study which would be ethically preferable.

Currently, investigators satisfy this requirement by considering whether it would be possible to collect the same or similar data by doing their research in test tubes or computers, rather than animals. Current regulations and guidelines do not, in contrast, require the investigators to assess whether it would be better to conduct the study in human beings. Indeed, many regulations *prohibit* researchers from testing new interventions in humans until they have been first tested in one, and sometimes two, species of animals. This requirement is based on the assumption that humans have significantly greater moral status than animals; hence, it is better to conduct the initial, and typically riskiest studies with animals: "When a new drug or surgical technique is developed, society deems it unethical to use that drug or technique first in human beings because of the possibility that it would cause harm rather than good."[29]

The conclusion that there are no degrees of moral status suggests that this approach is not justified. We cannot determine whether it is better to conduct a net-risk study in human beings or animals by appeal to their level of moral status. Instead, we need to assess the comparative harms and benefits in each case, and the extent to which the study is consistent with proper respect for animals and humans, as individuals and as agents. Because these assessments will vary depending on the study in question, it will not be possible to define a general preference; hence, reviewers should make these decisions on a case-by-case basis. This reveals a third way in which regulations governing animal research should be revised in light of the conclusion that there are no degrees of moral status. They should be revised to require an explicit assessment of whether it would be better to conduct a given study in human beings rather than animals.

Most obviously, it will be important to assess the value of a given study if it is conducted with human beings rather than animals. Because most interventions and medicines are being developed for use in human beings, studies using human subjects typically provide more useful information than studies with animals. Moreover, the current requirement to test experimental treatments in animals before humans is intended to weed out dangerous interventions before they are given to human beings. While this approach offers some protection for human subjects, it involves abandoning interventions that are found to be unsafe for animals. The problem is that results in animals frequently do not predict how experimental treatments will affect human beings. Hence, the current approach of abandoning interventions that are unsafe for animals likely leads to the rejection of experimental treatments that would have been safe and effective for humans. Assessment of whether it is better to conduct a study with animals or humans provides a way to take this concern into account.

One of the biggest challenges for assessing whether a net-risk study should be conducted in humans or animals is the fact that many of the potential harms of research involve aversive experiences. When it comes to human beings, we can ask how much they are upset or how painful a given research intervention is, such as a lumbar puncture. Because we cannot communicate with animals in the same way, we need to try to gauge the level of suffering based on their actions and reactions.[30] In addition, the committees which review research with animals need to keep in mind the point made earlier that human beings tend to assume that individuals who are physically different do not suffer to the same extent. In particular, they should be aware of the possibility that this bias might lead to their downplaying the suffering of animal subjects.

At the same time, we should not assume that an intervention which would cause significant pain to human beings will necessarily have similar effects with animals. Most obviously, human beings can understand in ways that animals cannot. This can have important implications for the harms that might be caused by an experiment. The fact that humans understand can lead to their suffering *more* in some cases. For example, research with human beings involves explaining what the study involves and what it will be like to participate. That explanation can result in human subjects experiencing anticipatory stress and anxiety that would not occur in animals. Our understanding of an experiment can also lead to less suffering in some cases. While humans who are informed know what is coming, they also can be told how long it will last. Being placed in a narrow metal tube and subjected to loud noises can be terrifying for individuals who do not understand what is happening or how long it might last. This experience can be educational and even enjoyable for individuals who understand that they are undergoing an essentially risk-free, thirty-minute magnetic resonance imaging (MRI) scan.

Our understanding of a study can reduce the suffering it causes in another way, too. To consider an actual example, I once sat on a committee charged with evaluating the ethical acceptability of a new method to study pain. The study involved injecting capsaicin, the active ingredient in chili peppers, under the skin on subjects' forearms, which stimulated receptors in the arm, sending signals to the brain and producing sensations of pain. The goal of the research was to find ways to block the pain signal between the site of the injection and the brain. The review committee agreed that the research was valuable, but it was concerned that the pain experience might be too much for the subjects. To assess this concern, the committee mandated that two of the investigators try the method on themselves first, and report back. Here is what they reported: following almost immediately after the injection, there is an experience of "fist-clenching" pain. That lasts about five to ten seconds, followed by significant but tolerable pain for a minute, which tapers off over the next few minutes to throbbing pain and then barely noticeable pain.

When the investigators described these experiences, I was struck by the fact that, although they were describing a significant pain experience, they did not seem to have been particularly bothered by it. Instead, they seemed to find the experience almost enjoyable, or at least valuable. This makes sense given what we know about the experience of pain and the extent to which it leads to suffering. Experiencing the pain taught them something of personal interest and something that was important for achieving their goals of getting the study approved and conducting their research. This is not unlike the experience of many long-distance runners and mountain climbers. The activities themselves are associated with moderate to significant aversive experiential inputs: cramps, muscle strains, twisted ankles, difficulty catching one's breath, pain. But, in many cases, these inputs do not result in suffering. Instead, they become part of a valuable activity, of learning something, of accomplishing something valuable. In these cases, possession of superior cognitive capacities provides a reason to prefer conducting the studies in human beings rather than in rabbits. Even when the pain is the same in the two cases, it might produce significantly less suffering in human beings.

This example highlights how things change when we stop analyzing ethical challenges that involve tradeoffs between humans and animals in terms of levels of moral status and instead consider the individuals themselves, and how they will be affected. This approach is more complicated. It requires that we consider their capacities, histories, levels of understanding, and other potentially relevant properties. But this approach also brings with it greater insight and understanding, in the present case, of how our superior cognitive capacities are relevant to and influence the experience of participating in a research study. Once we recognize that there are no degrees of moral status, this understanding increases the chances that we will make the right choice and, thereby, do the right thing.

Another important difference between animal and human experiments is that the harms to human beings typically involve only the experiences that occur during the study and any later side effects. Consider a study that involves getting blood drawn once a day for a week. Human subjects come to the clinic in the morning and then go home, or go to school or work for the rest of the day. Animal subjects, in contrast, are caged between visits, frequently alone, in an environment that is nothing like the places where they live and feel comfortable. These additional harms provide another reason to prefer doing studies in humans.

To this point, we have been considering differences in the experiential interests of human and animal subjects. While they are important, a full assessment should also consider differences in the subjects' contribution interests. Someone who dives into a river to save a drowning child has done something heroic and, thereby, becomes a hero. In the context of medical research, our superior cognitive capacities enable us to go beyond being mere subjects to recognize and embrace our participation as making a contribution to a valuable project. When we do, these projects not only have value in and of themselves, they have value for us, representing valuable contributions. In a survey we conducted, adolescents indicated that they benefited from participating in studies that posed risks to them in order to collect information to benefit others.[31] In particular, many of the adolescents indicated that they felt "proud" to be contributing to medical research. Rabbits will not feel proud of being in medical experiments. And, even if rabbits do have some contribution interests, it seems implausible to think that their interests are advanced by our putting them in research to benefit us.

In section 6.1, we saw that animals are owed respect. One way to show respect for others is to offer them compensation for the harms we subject them to. Human research participants who agree to participate in net-risk research are frequently compensated, sometimes with payments in the thousands of dollars. In addition, it is widely agreed that when humans are harmed as a result of participating in research, they should receive compensation. This suggests that burdensome and painful animal experiments can be made ethically better by compensating the subjects.[32] We saw this possibility in section 6.1 with respect to the researcher who purchases two rabbits, but ultimately ends up using only one of them in her pain-inducing experiment. If she has only one indivisible treat to share, she should give it to the rabbit subjected to the experiment. The researcher cannot undo the pain caused by the experiment. And she cannot increase the amount of pleasure that will be experienced by whichever rabbit gets the treat. But the researcher can make it the case that the pleasure of eating the treat is experienced by the rabbit that was subjected to the pain of the experiment. Respect for individuals as individuals thus provides a moral reason to compensate animals for the harms to which we subject them for our own benefit in medical experiments. Future work should consider ways to incorporate this approach into practice.

Full assessment of whether it is better to conduct a study in humans or animals should also take into account appropriate respect for the agency of the subjects. With respect to human beings, this involves preferring to conduct research with competent adults rather than children and obtaining the individuals' informed consent prior to enrolling them. Future research will be needed to determine how this factor influences enrollment of animals in research. One challenge here is that the capacity for agency varies widely across animals. I argued in section 6.2 that animals may have some claim to respect for their agency. Short-term involvement in research seems potentially consistent with these claims, while long-term studies and enrolling the same animals in multiple studies over an extended period might be problematic. Moreover, some animals, such as chimpanzees, have significantly greater capacity for agency, and researchers are evaluating whether it makes sense to conduct their studies only in animal subjects who agree to participate.[33]

In summary, the conclusion that there are no degrees of moral status does not establish that it is unethical to enroll animals in research that will not benefit them. At the same time, it does suggest that guidelines and regulations for animal research should be revised in significant ways. We have discussed several of the most important areas in need of revision. First, regulations should adopt a limit on the level of aversive experiences to which animal subjects may be exposed. Second, it will be important to ensure that animals are enrolled in net-risk studies only when the value of the study justifies the harms it causes them.

Third, before approving studies with animals, review committees should consider whether it would be better to conduct them in human beings. Our possession of superior cognitive capacities implies that some studies will cause us more pain and suffering than they will cause animals; other studies will cause us less suffering. Moreover, human beings can consent and participating in valuable research can promote our contribution interests. These are important reasons to prefer enrolling humans rather than animals. Fourth, guidelines and regulations should consider the possibility of compensating animals for the pain and suffering they experience in research studies. Fifth, future research should consider to what extent obtaining the agreement (also called assent) of some animals, such as chimpanzees, might offer a way to respect their agency.

Finally, this section has focused on how we should treat animal subjects for their own sake. This leaves the impact that conducting experiments with animals has on the researchers, and on humans more generally. In particular, to what extent does conducting aversive experiments on captive and uncomprehending animals harm the researchers themselves?[34] A comprehensive approach to animal ethics should take this concern into account as well.

8.6. Humans Who Lack Superior Cognitive Capacities

Proponents of degrees of moral status argue that possession of one or more superior cognitive capacities (to a sufficient degree) endows individuals with higher or full moral status. Newborns do not possess the cited superior cognitive capacities to any extent, while fifty million adults have Alzheimer disease which, over time, robs them of the superior cognitive capacities they once had. Moreover, hundreds of millions of human beings with cognitive deficits do not possess the cited superior cognitive capacities beyond the threshold for higher moral status. Postulation of degrees of moral status thus suggests that hundreds of millions of human beings have only baseline moral status, the same moral status as rabbits.

Proponents of degrees of moral status respond to this concern in different ways. Some argue that being human itself endows individuals with higher moral status. While this approach grants higher moral status to all human beings, it seems arbitrary, and it is often regarded as a kind of prejudice, *speciesism*, which is akin to racism or sexism. In particular, there does not seem to be any reason to think that the species to which an individual belongs determines their level of moral status, independent of the properties the individual themselves possess.[35] Other proponents accept the conclusion that only individuals who possess the cited superior cognitive capacities to a sufficient extent (sometimes called "persons") have higher moral status. On this view, the protections afforded by higher moral status do not apply to children, adults with severe dementia, and individuals with cognitive deficits, in which case it might be acceptable to torture hundreds of millions of human beings when the benefits to others justify it. To say the least, this view is problematic.

To avoid these concerns, without endorsing speciesism, some commentators argue that individuals' moral status is determined, not by the level of the cited superior cognitive capacities they have at the time, but by the highest level they attain over the course of their lives. If, at some point, an individual possesses a narrative identity, say, then they have higher moral status for the entirety of their life.[36] The first problem with this response is that it seems arbitrary. The fact that an individual will develop the cited superior cognitive capacity in fifteen years, or the fact that they possessed them fifteen years ago, does not seem to provide reason to think that they have full moral status now. More importantly, some human beings never develop the cited superior cognitive capacities to a sufficient extent. Eva Kittay writes about her daughter who does not possess the cited cognitive capacities, such as the capacity to recognize the principles of morality.[37] This suggests, on the view that higher moral status traces to possession of the cited superior cognitive capacities, that Kittay's daughter lacks higher moral status and the protections it affords. This view thus places individuals like Kittay's daughter who have lifelong cognitive deficits in jeopardy.

Some defenders of degrees of moral status argue that Kittay's daughter gains higher moral status on other grounds. Carl Cohen argues that individuals with lifelong cognitive deficits are of a kind—human being—that has high moral status in virtue of the fact that average humans develop the cited superior cognitive capacities.[38] Jeremy Waldron argues that all human beings, unlike animals, have the potential to develop the cited superior cognitive capacities.[39]

The fact that an individual might have developed the cited superior cognitive capacities to a sufficient extent, but didn't, or the fact that average members of their species develop them does not seem to imply that the individual themselves possesses higher moral status. Specifically, a species is frequently defined as a group of individuals who can breed with each other, but not with individuals outside the group (other definitions cite similarities in individuals' genes or genealogy). On this definition, being a member of the species *Homo sapiens* implies that an individual can breed with other humans and cannot breed with rabbits. However, the fact that one individual can interbreed with another does not imply that the first individual inherits the level of moral status possessed by the second.[40] In the end, then, postulation of degrees of moral status does not offer adequate protection for hundreds of millions of human beings.

Some might worry that the same problem undermines the present conclusion that there are no degrees of moral status. I have argued that, even though our superior cognitive capacities do not increase the level of our moral status, they do imply that frequently we have more at stake than rabbits; hence, it is frequently more important to treat us better. Our possession of superior cognitive capacities implies, for example, that we lose significantly more when we are killed; hence, it is significantly worse to kill human beings than rabbits. However, if how we should treat others depends on what they have at stake, and individuals who lack the cited superior cognitive capacities frequently have less at stake, it seems to follow that we do not need to treat them as well. If inadequate protection for individuals like Kittay's daughter provide a reason to reject degrees of moral status, it seems to provide a reason to reject the present conclusion as well.

This much is clear: how we should treat human beings with significant cognitive deficits is challenging and not always obvious. Hence, this issue will not provide a definitive reason to endorse or reject degrees of moral status. With that said, the rest of this section argues that the present conclusion that there are no degrees of moral status offers greater protection for these individuals compared to the claim that there are degrees of moral status. This analysis thus provides insight into the implications of the present conclusion and, in the process, offers additional support for it.

Many of the ways to harm others involve causing them to have aversive experiences. This is important because rejection of degrees of moral status provides significantly greater protection in this regard. Rejection of degrees of moral status implies that the experiential interests of humans who do not possess the cited superior cognitive capacities to a sufficient degree, their interests in avoiding pain and suffering, experiencing love and friendship, joy and pleasure, count as much morally as the analogous interests of individuals who possess the cited superior cognitive capacities. Indeed, in some cases, rejection of degrees of moral status affords greater protection for the experiential interests of humans with cognitive deficits. As we saw in section 8.5, the understanding that comes with the cited superior cognitive capacities often mitigates the suffering that results from aversive experiences. Understanding the goal of a medical experiment, and how long it will last, can significantly reduce the suffering it engenders. In these cases, there is reason to enroll individuals who can understand rather than those who can't. Endorsement of degrees of moral status suggests the opposite: the pain and suffering of individuals with cognitive deficits matter less; hence, it is ethically better to enroll them in pain-inducing experiments. This difference provides further reason to reject degrees of moral status.

The other major ways of harming or wronging individuals that we have considered are failing to respect them and killing them. If there are no degrees of moral status, the ethics of killing individuals is determined by their moral action guiding properties. I have argued that the death of human beings is problematic, not because of their higher moral status, and not because of the future positive experiences it deprives them of, but because it ends their existing relationships and projects, as well as the potential for future ones.

Proponents regard the protections afforded by higher moral status as all or none. Individuals who possess the cited superior cognitive capacities (e.g., autonomy or moral agency) to a sufficient extent enjoy all the protections that higher or full moral status affords. All other humans, several hundred million of them, do not enjoy any of these protections. Rejection of degrees of moral status offers a very different picture, and greater protections in this regard as well. The fact that an individual does not possess the cited superior cognitive capacities to a sufficient degree is consistent with their possessing a wide range of cognitive capacities. It is even consistent with their possessing the cited superior cognitive capacities to some extent.

The cognitive capacities that these individuals do possess frequently enable them to make contributions to valuable projects and relationships. This possibility was illustrated by the woman we discussed in section 6.2 who had lifelong cognitive impairments. If there are degrees of moral status, she would not qualify for full moral status. Still, she was able to understand a good deal about the proposed research project and, based on that understanding, she embraced the prospect of participating in and contributing to it, to research to find treatments for her disease. Similarly, many humans who do not possess the cited superior cognitive capacities to a sufficient degree still embrace and contribute to rich relationships with their friends and families. Rejection of degrees of moral status and focus on individuals' moral action guiding properties, rather than whether they have full moral status, reveals that death is bad for these individuals, for their own sakes. The present conclusion thus implies that there are important moral reasons not to kill them.

This leaves respect for the agency of humans with cognitive impairments. On the view that there is a threshold on possession of the cited superior cognitive capacities beyond which individuals have full moral status, the fact that some human beings fall below the postulated threshold implies that there are no (non-welfare) reasons to respect their agency. Rejection of degrees of moral status, in contrast, is based on recognition of the fact that the types of thresholds required by this view don't exist. Without them, the importance of respect for individuals' agency depends on the extent to which they possess the capacities for agency and the extent to which exercising them is an important part of a flourishing life for the individuals. This view suggests that there *are* reasons to respect the agency of human beings who have cognitive deficits. At the same time, the reasons to respect their agency are weaker than the reasons to respect the agency of average human beings. Is that problematic?

The extent to which we should permit individuals to lead their own lives, versus giving them more guidance and being less hands-off, depends on the extent of their abilities in this regard. For example, as children mature, and their cognitive capacities increase, they go from having essentially no leeway to make their own decisions, to increasingly more. This same approach was illustrated by the woman discussed in section 6.2. Her impairments suggested that she should not be given as much leeway to make her own decisions and take on research risks as competent adults. But the fact that she understood important aspects of the study and wanted to participate based on her own values provided reason to let her make this decision.

The present discussion highlights important differences between endorsement of degrees of moral status and rejection of them. Endorsement places the focus of ethical analysis on individuals' level of moral status. On this approach, the superior cognitive capacities that are thought to provide the basis for greater moral status are the primary properties which determine how we ought to treat others, for example, whether we ought to allow them to make their own decisions. This is why endorsement of degrees of moral status simplifies the moral life. Whether we have reason to respect some individuals' agency and to what extent can be determined simply by determining whether they have full moral status.

This same approach threatens human beings who do not possess the cited superior cognitive capacities to a sufficient degree. They are in danger of being regarded as having lower moral status to the extent that it is appropriate to treat them like rabbits. Importantly, this concern arises only if there are degrees of moral status, including levels below our own. Rejection of degrees of moral status thereby offers a definitive response to this worry: humans with cognitive deficits cannot occupy a lower level of moral status for the simple reason that there aren't any; there is only one level of moral status.

The conclusion that there are no degrees of moral status replaces the almost singular focus on whether individuals possess the cited superior cognitive capacities to a sufficient extent with a broader assessment and appreciation of individuals and the capacities they do possess. This is the point Eva Kittay is making when she explains that her daughter, who lacks the superior cognitive capacities which are cited as the basis for higher moral status, has enormous capacity to love and to bring joy into the world, to enter into meaningful and deep relationships with others. And to experience pain, and loneliness, and to suffer. Rejection of degrees of moral status provides the opportunity to consider how these capacities give her daughter important claims to our protection and our respect. Rejection of degrees of moral status thereby explains why we should not treat humans who lack the cited superior cognitive capacities like rabbits, nor rabbits like humans who lack superior cognitive capacities.

Consider an example which illustrates this difference in approaches to how we should treat others. Individuals with Alzheimer disease frequently lose their short-term memories before they lose their long-term memories. As a result, many individuals with Alzheimer disease remember that they were married, but they forget that their spouse recently died. When they are reminded, the individual might be distraught for a day. The next day, though, they have forgotten and are comforted by the belief that they are still married. What should caregivers do the next time the individual asks about their spouse? Should the caregivers tell the truth, even though it will subject the individual to significant distress, over and over again, every day? Or should the caregivers lie?

Postulation of degrees of moral status offers a clear answer. If all human beings have higher moral status, lying to these individuals is unethical. Like all humans, they have a right to be told the truth, even when they find the truth distressing. Alternatively, if humans with cognitive deficits do not have higher moral status, lying to them is perfectly fine. Like rabbits, they have no right to the truth. Endorsement of degrees of moral status thus has the virtue of offering a clear answer to an important ethical question: simply look to the level of moral status possessed by individuals with cognitive deficits. Nonetheless, that strikes me as a mistake. To figure out how caregivers should respond, we need to look to the capacities these individuals possess and consider what proper treatment of individuals with those capacities involves. In other words, don't ask what proper treatment of individuals with a certain level of moral status involves. Ask what proper treatment of these individuals involves. That is the approach which rejection of degrees of moral status recommends. It strikes me as the right one, which provides further support for the conclusion that there are no degrees of moral status.

8.7. Superbeings

Discussion of degrees of moral status frequently focuses on two questions: whether animals have lower moral status than human beings, and whether all human beings have the same moral status. We have largely followed this approach, assessing whether there are one or more levels of moral status *below* our own. At the same time, the claim that there are different levels of moral status raises the possibility that there might be levels above ours. In particular, proponents claim that we have significantly higher moral status than animals because we possess superior cognitive capacities. This suggests that individuals whose cognitive capacities are superior to ours, what are sometimes referred to as "superbeings," could have higher moral status than we do. If we have higher moral status than animals because we have significantly greater capacity for moral agency, for example, individuals who are significantly more capable moral agents than we are would be superbeings, and they would have greater moral status than we do. While postulation of degrees of moral status offers a way to establish that we are more important morally than animals, it thus threatens us with lower moral status than superbeings.

If our eating animals and using them in pain-inducing experiments is justified by the fact that we possess a particular cognitive capacity to a higher degree, it seems to follow that superbeings whose cognitive capacities are significantly greater than our own would be justified in eating us for dinner and using us in their pain-inducing experiments. Moreover, if superbeings' moral status is higher than ours, this might not be a matter of their acting unethically for their own benefit; it might be precisely how they ought to act. They ought, for example, to perform pain-inducing experiments on us when doing so offers a chance to find treatments and cures for diseases that afflict their superior selves. And resistance on our part might be worse than futile, it might be unethical. We might be morally obligated to facilitate their using us for their benefit.

There are at least three possible sources of superbeings. First, there might currently be intelligent beings in other corners of the universe whose cognitive capacities exceed our own. Second, our traits, including our cognitive capacities, are a result of the interactive and cumulative effects of our environment and genes. Changes to our environment and/or our genes thus have the potential to significantly increase our cognitive capacities. This has always been a topic for science fiction, but the science is starting to catch up to the fiction.

Most notably, scientists have identified a method, called CRISPR, that allows them to edit individual genes. As this technology advances, and even more advanced ones are developed, scientists might be able to create human beings who are cognitively superior to the rest of us.[41] Average human beings are able to choose pretty effectively between three and four options. When presented with more options, our ability to make decisions quickly breaks down. We get confused and tend to revert to simple heuristics: choose the first one or the most prominent one. Human beings are not very good at reasoning in response to probabilities either. We tend to focus on low and high probabilities, treating all the others as pretty much the same. Genetic enhancement might produce humans who are able to choose effectively among twenty options, or two hundred, to make decisions that distinguish between a 20 and 30 percent chance, or even between a 20 and 22 percent chance. Genetically enhanced humans might be:

> consistently impartial whenever impartiality is morally required. Second, because they screen out distracting stimuli and think very quickly, they reach correct moral judgements in conditions of stress no less consistently than they do in leisurely reflection.

> Third, they suffer from weakness of will so seldom that any of their members who does so is regarded as having a psychological disorder. Finally, in comparison with ordinary persons, these enhanced humans are enormously adept at envisaging the likely consequences of their choices and identifying the implications of their moral judgments.[42]

In section 7.2 we considered the fact that parents should be partial to their own children and help them to lead better lives, even when other children need their help as much, or more. This is important because having greater cognitive capacities and correspondingly higher moral status could benefit children in many ways. It would put them first in line for life-saving organs and other scarce resources, it would make their pleasure more important ethically, and their pain ethically worse. Future trillionaires might thus conclude that ethically they should spend their money to pay for genetic manipulations to endow their children with enhanced superior cognitive capacities and, thereby, give them higher moral status than children who don't happen to be born to super-rich parents.

A third source of potential superbeings comes from computers and robots, which already have greater capacities than human beings in several respects. They are better at diagnosing certain diseases and predicting disease outcomes than physicians,[43] their memories are vastly superior, and they are better at chess. It is only a matter of time before advances in machine learning and artificial intelligence produce computers and robots whose cognitive capacities are superior to ours in essentially every respect. And, if sentience is necessary for moral status, they might develop that too.[44] At that point, if there are degrees of moral status, they would have higher moral status than everyone else, except perhaps the children of the super-rich, no matter which property or properties one assumes provide the basis for higher moral status.[45] This would place us in significant jeopardy relative to superbeings.

One might hope that it is not possible for superbeings to become significantly more important simply because we already occupy the highest level of moral status. This possibility is suggested by the claim, frequently endorsed by proponents of degrees of moral status, that we have "full" moral status. This claim would make sense if there were a scale for the moral importance of individuals' rights, interests, or claims which has a maximum value, say, from 0 to 100. If that were the case, and there were reasons to believe that we occupy a very high position on the scale, perhaps 97, it would not be possible for superbeings to gain significantly higher moral status than (unenhanced) human beings. Unfortunately, this is not the way moral status works.

Levels of moral status are *comparative*: having higher moral status than rabbits means that our interests are more important, our rights are stronger, than the analogous interests and rights of rabbits. Superbeings could thus have higher moral status than we do in the sense that their rights, interests, or claims are more important morally than ours. In that case, when there is a conflict between respecting the rights or protecting the interests of us versus respecting the rights or protecting the interests of superbeings, the morally right thing to do would be to protect or respect theirs rather than ours. This suggests that, if our superior cognitive capacities can increase the level of our moral status, there could be levels of moral status above our own and, eventually, superbeings who occupy those levels.

Individuals who are sentient have moral status, which involves their interests mattering morally. This is the first level of moral status. Some commentators argue that sentient individuals who are moral agents also have rights. We can think of this as the second level of moral status, the one that we are assumed to occupy. In principle, there could be a third, and perhaps even a fourth, level of moral status which involves the rights of individuals being stronger or having more moral weight than our rights.[46]

This would not be possible if the rights of individuals at the second level, our level, are absolute.[47] If, for example, our right to direct the course of our lives were absolute in the sense that it is unethical for other moral agents to ever interfere with our decisions about what we do and how our lives go. Yet our rights are not absolute. It is appropriate for you to interfere in minor ways with my right to direct the course of my life when doing so yields significant protection or benefits for others. It is appropriate for you to prevent me from crossing the street for a minute when doing so is necessary to save another person from serious harm. The higher moral status of superbeings, and their correspondingly stronger right to direct the course of their lives, might imply that it is permissible to stop them from crossing the street for a minute only when doing so offers the chance to save one hundred human beings from serious harm.

It also is permissible to harm one human being *significantly* when doing so is necessary to avoid dramatic harm to many others. It is, for example, ethically acceptable to act in a way that results in one person losing several fingers when doing so is necessary to save the lives of one hundred people. In contrast, it might be permissible to do this to a superbeing only when it is necessary to save the lives of one thousand human beings. Or a superbeing's moral status might be so much greater than ours that it is never acceptable to harm a superbeing in order to save any number of human beings. Perhaps superbeings may be harmed intentionally only when it is necessary to save other superbeings.

As we have seen, the conclusion that there are no degrees of moral status complicates the moral life in a number of ways. But in this case at least, it makes things less complicated. The fact that there are no degrees of moral status implies that, no matter how smart they become, superbeings cannot gain greater moral status than we have for the simple reason that there are no degrees of moral status among those who matter morally. This conclusion addresses many of the concerns that the potential for superbeings raises. For example, we need not worry about being justly enslaved by superbeings for the simple reason that they cannot have significantly higher moral status, which would trump our claims to direct the course of our own lives.

The present conclusion similarly addresses a number of ethical concerns that are discussed with respect to the development of so-called animal-human chimeras. Experiments which inject human cells into the embryos of animals provide the opportunity to develop animals with human characteristics and thereby study the development of human cells. Scientists have already used this process to create pig-human embryos and chicken-human embryos.[48] In one case, the Japanese government gave scientists permission to try to grow animals with human organs. The experiment basically involves deleting the genes responsible for developing particular organs in the animal embryo and inserting genes that direct the development of a human version of the organs. If successful, this research might offer a source of organs for the many human beings who are on waiting lists and the many people who die before one becomes available. In principle, this line of research could lead to as many human organs as we need, potentially saving thousands of lives every year. This line of research also raises important ethical issues.

One issue is whether it is permissible to manipulate the genes and the anatomies of animals in radical ways. Is it, in effect, permissible to create new types of animals? The conclusion that there are no degrees of moral status—hence, animals' moral status is similar to our own—makes these questions difficult to answer. We cannot defend this research by arguing that our significantly higher moral status justifies exposing animals to essentially any risks and harms for our benefit. Instead, we will need to estimate the harms to them, the potential benefits to us, and compare them. We will also need to consider to what extent these experiments involve treating the animals with appropriate respect.

In another way, the present conclusion makes the ethics of these experiments much easier. Many commentators express concern that manipulating animals in ways that increase their moral status is itself unethical. This suggests that, prior to permitting these experiments, it is critical to determine the properties which are the basis for our higher moral status and make sure that scientists do not endow animals with those properties. If our greater moral status traces to one or more superior cognitive capacities, for example, it would be problematic to use these techniques to endow animals with similar cognitive capacities. With this in mind, the Japanese government stipulated that the researchers must halt their experiments if they find any human cells in the animals' brains. To the extent that this restriction traces to concerns about increasing the moral status of the animals, the present analysis reveals that it is unnecessary. It is not possible for this, or any other experiment to give sentient animals the same level of moral status as human beings. They already have it.

Summary

The present chapter considered the implications of the conclusion that there are no degrees of moral status for seven important issues. Section 8.1 considered how this conclusion changes things and how it doesn't. It doesn't suggest that we should treat animals like humans, or humans like animals. Instead, it suggests that, in determining how we ought to treat others, we need to look to their moral action guiding properties, as opposed to their level of moral status. This approach understands the moral significance of our superior cognitive capacities very differently: rather than increasing our level of moral status, they frequently imply that we have more at stake compared to animals.

In section 8.2, we saw that the present conclusion does not imply that we have to spend all our time helping animals. The reason is that the principle of prioritizing the worse off concerns the extent to which individuals are living flourishing lives, which depends on the type of individual they are. Sloths might well be living flourishing lives, even if they are not living particularly rich ones.

Section 8.3 discussed the fact that sentient animals have the same rights against being harmed for the benefit of others. In contrast, whether they have a right to direct the course of their lives, and a right against being killed, depends on whether they have the requisite capacities, and the extent to which directing the course of their lives is part of a flourishing life for them (independent of the extent to which it promotes their experiential interests). This seems unlikely for rabbits and many animals, although it might be the case for others.

Section 8.4 noted that the present conclusion establishes that raising animals for food in ways that cause them significant suffering is unethical. This led to the question of whether humane animal farming might be acceptable. I argued that humane farming is ethical with respect to the impact it has on animals who are minimal agents, but not on animals with more sophisticated cognitive capacities, such as pigs and octopuses. This suggests that such practices are acceptable in some animals, such as chickens and rabbits, pending further assessment of the impact the practice has on the humans engaged in it, and whether it is consistent with our having appropriate relationships with the animals.

The conclusion that there are no degrees of moral status revealed, in section 8.5, that regulations governing animal research should be revised in several important ways. They should place an upper limit on the extent to which animal subjects may be exposed to aversive experiences and mandate that the potential benefits of the research must justify the harms to the subjects. They should also mandate assessment of whether it would be ethically preferable to conduct a given study in human rather than animal subjects. Finally, it will be important to consider whether it would be feasible to compensate animals for the harms they experience as research subjects and whether it makes sense to solicit the agreement (assent) of some animal subjects.

Section 8.6 discussed the implications of the conclusion that there are no degrees of moral status for human beings who do not possess the cited superior cognitive capacities to a sufficient degree. In particular, we saw that it likely offers them greater protection than views which endorse degrees of moral status. Section 8.7 pointed out that we need not worry about superbeings' moral status exceeding our own for the simple reason that there are no levels of moral status above (or below) our own. For the same reason, we need not worry that experiments which involve placing human cells into animals might increase the animals' level of moral status.

Conclusion

A Future Without Hierarchies

C.1. Summary of the Argument

Many people believe there are degrees of moral status, a hierarchy among individuals who matter morally. This view is supported by our intuitions in response to specific cases, such as *Experimentation* and *House on Fire*. *Experimentation* suggests that it is worse to cause human beings to suffer than it is to cause rabbits to suffer, while *House on Fire* suggests that it is worse for human beings to die. At the same time, not everyone shares these intuitions. Some people think there are no degrees of moral status, and rabbits are our moral equals. They think it is no better morally to conduct pain-inducing experiments with rabbits than humans. To try to resolve this disagreement, we have gone beyond intuitions to systematically assess whether there are degrees of moral status.

After setting the background in Chapter 1, this assessment began in Chapter 2, with a discussion of why we cannot dismiss degrees of moral status out of hand, followed by a description of three types of morally relevant properties: moral status conferring properties, moral action guiding properties, and moral status enhancing properties. Moral status conferring properties determine whether individuals possess moral status at all. Moral action guiding properties determine how individuals who have moral status ought to be treated. However, as we saw, moral action guiding properties cannot provide a basis for degrees of moral status, leaving the possibility that one or more properties are moral status enhancing;

Life Without Degrees of Moral Status. David S. Wendler, Oxford University Press.
DOI: 10.1093/oso/9780197675328.003.0010

that is, they increase the moral status of the individuals who possess them. For example, some people think that sentience gives individuals moral status while possession of one or more superior cognitive capacities gives sentient individuals higher or full moral status.

In Chapter 3, we assessed whether any properties might be moral status enhancing, concluding, for several reasons, that none are. First, animals possess many impressive properties. Some fly, others are extremely selfless, working and sacrificing themselves for others, and some might even be immortal. If possession of impressive, even transcendent and morally relevant properties increased the moral status of those who possess them, these animals would have significantly higher moral status. The fact that they don't suggests no properties are moral status enhancing.

Second, human beings possess many negative capacities, including the capacities for greed, selfishness, cruelty, and genocide. If possession of some properties could influence the moral status of those who possess them, these properties would diminish our moral status; they might imply that we have lower moral status than individuals who are otherwise similar but do not possess these capacities. Our possession of these capacities might even imply that we have negative moral status. The fact that these properties don't diminish our moral status, and we definitely do not have negative moral status, suggests that no properties influence the moral status of those who possess them, supporting the conclusion that there are no moral status enhancing properties.

Third, in addition to doing many wonderful things, human beings have done many terrible things. If the properties we possess could influence our level of moral status, this history could lower ours. The fact that it doesn't further supports the conclusion that no properties influence the level of individuals' moral status. Moreover, our superior cognitive capacities enable us to do terrible things, to plan and carry out genocide, for example. This raised the question of whether it even makes sense to characterize our superior cognitive capacities as morally positive, hence, potentially moral status enhancing. Perhaps the fact that these capacities put us in position to do wonderful and terrible things implies that they are positive and negative, or that it does not make sense to regard them as positive or negative. This possibility further undermines postulation of degrees of moral status.

Fourth, moral action guiding properties are relevant to specific actions or ways of treating others. The capacity to experience pain is relevant to the ethics of sticking rabbits with needles as part of medical experiments, but not to the ethics of killing them painlessly or taunting them for losing races to turtles. Moral status enhancing properties, if they existed, would be different. They would imply that it is worse to treat individuals negatively even when possession of the property in question does not increase what the individual has at stake. It would be worse, proponents argue, to cause a given amount of pain and suffering to human beings than it would be to cause the *same* level of pain and suffering to rabbits. This aspect of moral status enhancing properties raises the problem of relevance: How could possession of some properties make it worse to treat individuals negatively when the property does not increase what they have at stake? How could possession of some properties make it the case that individuals' interest in not suffering matters more or that their claims to be treated with respect are stronger?

The answer offered by proponents of degrees of moral status is that possession of one or more properties makes the individuals who possess them morally more important, which is understood in terms of the individuals' interests or claims mattering more. The problem is that appeal to individuals being morally more important is supposed to explain why their interests or claims matter more than the analogous interests or claims of others. By defining greater moral importance in terms of one's interests or claims mattering more, proponents fail to address the problem of relevance. Rather than explaining why possession of one or more properties makes individuals' interests or claims matter more, they adopt another term, higher moral status, to repeat that they do. This circular reasoning confirms the absence of moral status enhancing properties.

Fifth, and finally, analysis of the nature of thresholds revealed why no properties are moral status enhancing. All the plausible candidates for moral status enhancing properties come in degrees. Endorsement of degrees of moral status thus suggests that there are countless levels, with essentially every increase, no matter how small, in one or more moral status enhancing properties yielding greater moral status. However, there are not countless degrees of moral status. There are not, for example, hundreds of millions of levels of moral status among human beings.

Postulation of countless levels of moral status also magnifies the problem of relevance. Proponents who endorse a few levels argue that possession of one or more superior cognitive capacities increases individuals' moral status as a result of making the individuals themselves morally more important. While we ultimately rejected that view, it makes sense. For example, it makes sense to claim that individuals who are moral agents or autonomous are more important than individuals who lack these capacities. In contrast, postulation of countless levels of moral status implies that the smallest possible increase in one or more moral status enhancing properties makes individuals morally more important. That does not make sense. It does not make sense to claim that your pain matters more than my pain, and you get priority for life-saving organs, because you are able to remember the past to the smallest degree better than I can.

To avoid these concerns, most proponents endorse thresholds on the cited moral status enhancing properties, which account for the fact that not every increase yields higher moral status. Instead, only increases that take the individual across the threshold yield greater moral status. If there are two levels of moral status, proponents need to endorse only one threshold; if there are a few levels, they need to endorse a few. This response to the problems raised by countless levels of moral status led to an exploration of the nature of thresholds. We saw that there are non-arbitrary thresholds of the type that would be required for levels of moral status, focusing on the boiling point of water. However, these thresholds involve competing factors, with the threshold occurring at the point at which one factor exceeds the other. This insight suggested that the possibility of degrees of moral status depends on the existence of a competing force to moral status enhancing properties such that the threshold occurs where the moral status enhancing properties exceed the moral status decreasing properties. The possibility of degrees of moral status thus depends on the existence of moral status *decreasing* properties. The fact that moral status decreasing properties don't exist revealed that there are no non-arbitrary thresholds which could provide the basis for degrees of moral status; hence, there are no degrees of moral status.

In light of this conclusion, we reconsidered the three examples of significant harms or wrongs that we encountered in Chapter 1. In Chapter 4, we saw that it is worse to kill human beings than it is to kill animals. However, this is explained, not by a difference in our level of moral status, but by the fact that killing a human being sets back their contribution interests; most animals have minimal, if any, contribution interests. Killing human beings also violates their strong claim to direct the course of their own lives; most animals have no or significantly weaker claims in this regard.

In Chapter 5, we considered whether it is worse for humans to suffer than for animals to suffer. We have seen that, although many of us have intuitions that it is, our intuitions in this respect are not reliable indicators of the truth. We tend to assume that pain and suffering are less bad when they occur in individuals who are significantly different than we are. Moreover, we tend to assume that individuals who are physically different from us have lower moral status, in which case our intuitions about the relative moral importance of different instances of pain and suffering might be a result of, as opposed to providing support for, the belief that we are morally more important than animals. For these reasons, consideration of the ethics of pain and suffering, and the intuitions they prompt, do not provide evidence for or against degrees of moral status.

Chapter 6 considered whether respect really applies to human beings, but not animals, focusing on respect for individuals as individuals and respect for agency. We identified cases in which there are reasons to respect animals as individuals and possibly reasons to respect their agency. To assess these intuitions, we briefly considered the nature of respect. This analysis suggested that individuals' claims to be treated with respect trace to the unique subjective existence of sentient individuals. It follows that there are moral reasons to respect sentient individuals, even when they do not possess superior cognitive capacities, such as autonomy, a rich narrative identity, or even self-consciousness.

Appropriate respect is a matter of degree, where the strength of the reasons to respect others increases with the complexity of their subjective existence, the complexity of their agency, and the extent to which respecting their agency is an important component of a flourishing life for them. These factors suggest that it is more important to respect human beings, but there are moral reasons to respect sentient animals.

Chapter 7 considered two pragmatic arguments that we should believe in degrees of moral status despite the conclusion that they don't exist. In section 7.1, we considered the possibility that believing in degrees of moral status is a valuable rule of thumb: Given that human beings often have more at stake, believing that human beings have higher moral status might make it more likely that we do the right thing. We rejected this possibility on the grounds that this belief is more likely to result in our acting inappropriately. This was supported by the consequences of the historically prevalent belief that we are more important than animals, leading us, in section 7.2, to consider whether belief in degrees of moral status is supported by proper partiality, analogous to the widely endorsed claim that parents should favor their children over other children. Specifically, we considered whether we should favor other human beings over rabbits because other humans are our fellows.

The conditions for proper partiality—namely, that others also benefit sufficiently by having individuals who favor them—do not apply in the case of at least many animals. Moreover, proper partiality does not determine who we should favor, and it does not support many of the actions which are thought to be supported by degrees of moral status, such as experimenting on animals for our benefit. This chapter thus concluded that, not only are there no degrees of moral status, but we should accept this conclusion and focus on identifying it implications. Chapter 8 thus considered the implications for seven important issues. With that analysis in place, the next and last section, C.2, begins to explore a future without moral hierarchies, considering what we need to learn to figure out how to live ethically in a world without degrees of moral status.

C.2. Implications of the Conclusion

The present discussion reveals that there are no moral status enhancing properties. And, since degrees of moral status depend on the existence of such properties, it follows that there are no degrees of moral status. You either have moral status or you don't. Of course, not everyone will immediately agree. To recall the history of theories of the physical universe, when scientists started noticing things that conflicted with the claim that we inhabit its center, they responded by describing increasingly complex theories which were designed to explain what they saw. For any new piece of evidence, clever scientists were able to describe ever more complicated movements of the planets that were consistent with them.

This response was supported by several factors. First, people were used to the theory that the Earth is located at the center of the universe, and they were comfortable with it. Second, it made sense of our ordinary experiences, the sun rising in the east and setting in the west, the lack of wind on many days. Third, it has the reassuring consequence of placing us at the center of the universe. What brought this process to a halt was not the exhaustion of scientists' creative powers to describe ever more complicated theories, but the description of an alternative view: the sun is at the center of our universe. Initially, scientists did not embrace this view because it answered all their questions. To the contrary, many questions and concerns regarding this view were answered only after scientists began to take it seriously, after they began to accept that this is the world we live in and sought to understand it. And a number of important questions remain to be answered.

The present conclusion that there are no moral status enhancing properties—hence, no degrees of moral status—brings us to a similar place with respect to the moral universe. Many people likely will continue to endorse the view that we are more important morally than animals.[1] We are used to this view. It has a long history. It has the reassuring consequence of placing us at the center of the moral universe. This view is also consistent with the intuitions many of us have in response to cases like *Experimentation* and *House on Fire*. And, for many people, believing that we are morally more important than animals provides a reason to feel good about themselves: whatever else might be true, at least we aren't animals.

To this extent, the conclusion that animals are our moral equals can feel threatening, similar to the ways in which the conclusion that we aren't more important than humans who don't look like us or have different hometowns can feel threatening. It eliminates our presumed comparative advantage and might thus be experienced as a loss. The way to address these concerns isn't to offer additional arguments in favor of the present conclusion. It's to explore and show that we can understand and, even flourish, in a world without degrees of moral status.

The conclusion that there are no degrees of moral status—hence, all individuals who possesses moral status are moral equals—does not reveal how we should treat others; it does not reveal how we should treat rabbits, genius robots, or animal-human chimeras. It does not, for example, instantly reveal which animals merit respect for their agency and to what extent. What, then, will a future without moral hierarchies be like?

In some contexts, and for some purposes, the differences are significant; for others, little, if anything, will change. It is somewhat like learning that the Earth is not the stationary center of the universe. That realization changed everything and it didn't. It meant that we did not occupy the center of the universe and that raised questions about the importance of us and our planet relative to everything and everywhere else. Maybe the Earth is not more important than other planets. Those implications were terrifying to many people, and individuals who claimed that the Earth is not at the center of the universe were scorned, imprisoned, even killed, as a result. Still, the fact that we inhabit a planet that is constantly rotating and revolving does not undermine our ability to raise crops to eat, build houses to live in, and go for walks without falling off. So we adjusted, and now it seems hard to imagine why people were so worried in the first place.

Our response to the present conclusion should be similar. Even without degrees of moral status, it is significantly worse to kill human beings than it is to kill rabbits. And, ethically, it is important to respect human beings, to respect them as individuals and to respect their agency, to help them to lead flourishing lives. These are some of the most important aspects of the moral life, and the present conclusion does not change how we should think about them.

The present conclusion also suggests that the ethics of some issues are more straightforward than many people have supposed. Concern has been expressed over the possibility that the future development of sentient robots or enhanced humans might lead to their having greater moral status than we do. And it has been thought that introducing human cells into animals, creating animal-human chimeras, could increase their moral status to the level of human beings. While these possible developments raise important issues, they are easier in this sense: we do not have to worry that they might increase the moral status of the beings in question for the simple reason that there are no degrees of moral status. The fact that all sentient individuals have the same level of moral status reveals that the only way to change individuals' moral status is to give sentience to those who lack it, or to take sentience away from those who possess it.

Most of us have lived our lives with the belief that, morally, we are the most important beings in the world. Our views on how we should treat others have thus been shaped by the implicit assumption that the other, the one we want to treat ethically, is a human being. As a result, our understanding of moral principles often implicitly concerns how we should treat humans. We believe, as a result, that it is terrible to kill others, to sterilize them, to keep them in cages, to have them eat off the floor. To understand a world without degrees of moral status, we will need to reconsider these beliefs: To what extent do they reflect how we should treat any individual who has moral status versus to what extent do they reflect how we should treat human beings?

The conclusion that there are no degrees of moral status does not imply that animals should be treated like human beings, or that human beings should be treated like animals. It also does not imply that sentient animals have the same fundamental rights as human beings. Instead, it implies that they have the same moral status as we do. Their pain and suffering matter as much as ours. And they deserve respect as unique subjective existences. If human beings did have significantly greater moral status than animals, we could often determine how we should treat animals simply by appeal to their level of moral status. We matter much more than they do; hence, in cases of conflict between us and them, we should favor us. And the fact that they matter so much less suggests their interests don't matter all that much, at least when they conflict with ours.

Life without degrees of moral status is very different. The fact that an individual has moral status tells us only that it matters morally. It does not tell us how we ought to treat it. For that, we cannot appeal to their level of moral status; we need to look to their moral action guiding properties. This raises fundamental questions regarding how we determine the moral action guiding properties of individuals. Pain and suffering are easy because it is intrinsically bad for individuals to suffer. But how do we determine whether other ways of treating individuals are morally appropriate? Put differently, how do we determine whether a given capacity of an individual is moral action guiding and to what extent?

Consider the fact that both rabbits and human beings have the capacity to choose mates and to procreate. It is important for human beings to be able to choose their own mates and to decide whether to procreate. One might be tempted to conclude from the fact that animals have the same moral status that it is also very important for them to choose their own mates and to decide whether to procreate. And that would imply, morally, that it is important for us to permit them to do so. This would follow if the moral importance of a given capacity follows from one's moral status. Or if the moral importance of our choosing mates for ourselves and deciding whether to procreate follows from the importance of these things for all individuals who have moral status. However, the fact that it is important for us to choose our mates, and decide whether to procreate, traces to what counts as a flourishing life for human beings. To determine whether these things are important for rabbits, we need to determine what counts as a flourishing life for them.

To return briefly to an example with which we started, is it appropriate to push the cat off the sofa and give the spot to a human being? Degrees of moral status make questions like this one easy to answer. Without them, it seems like we will have to consider myriad issues relevant to whether it is better for a human being or a cat to sit on the sofa. In fact, I don't think we need to do that in this case. We have good reason, independent of our being morally more important, to keep the spot for the human being. It is a party for human beings, after all, and we have good reasons to think that the cat will be fine.

Of course, we could be wrong. Our insight into the experiences of cats, and our ability to communicate with them, is limited. We thus might be wrong about what is it like for the cat to lie in the corner during parties. For the dog to eat off the floor. For the horse to be ridden through the fields. For the fish to live in a tank in someone's living room. The fact that these questions are critical to what we should do, together with the fact that the answers are not entirely clear, reveals that the moral life is more complicated and more uncertain than many have thought and most of us would like it to be. To some extent, this will be unsettling. It is morally important to do the right thing, but, in the end, and in many cases, we can't be sure what that is. The present conclusion leaves us with the challenge of coming to terms with a future in which there is much more uncertainty.

We did not choose to share the world with beings whose sentience gives them moral status. Nor did we choose powers of insight and communication which are insufficient to determine what is good for them, what constitutes a flourishing life for them. The conclusion that there are no degrees of moral status thus suggests that we will need to be more humble, and more forgiving. This might be the biggest challenge we face in a future without moral hierarchies. We do not deal well with uncertainty. And we often respond to it by imagining it is not there, that we can be certain what is right and what is wrong. A future without moral hierarchies poses the challenge of continuing to try to do the right thing, while recognizing that we can't be sure what that is and we might often fail despite trying our best.

These implications largely concern the experiential interests of animals. However, we have seen that there are also reasons to respect animals as individuals and possibly to respect their agency. Unfortunately, the conclusion that there are no degrees of moral status does not magically reveal how to do that either. Consider individuals who train horses to the point where the horse responds to commands almost without the rider ever giving them. After several years, the horse and rider win an Olympic medal, an amazing achievement. To assess whether this involves appropriate respect for the horse, the natural response is to appeal to models of relationships we know and understand: friend, neighbor, teammate. Some might thus compare this to a sports team which practices together for years, eventually winning important competitions. Viewed in that light, riding horses and training them to win medals seems more than acceptable; it seems impressive. But others might compare this case to a slave owner who locks his slave in the barn every night and brings him out in the morning, training the slave to carry the owner on his back, whipping him when he stumbles. This suggests, not surprisingly, that treating horses in this way is abominable.

Our responses to both analogies depend critically on the fact that the individuals in them are human beings. It is wonderful for a team of humans to work together and win a gold medal. It is terrible for human beings to be locked away at night and told how to behave during the day. But it's not clear that these examples tell us anything about what respect involves when it comes to horses. Is it better for horses to roam the prairies or to compete in the Olympics? Is it better for them to sleep in the wild or in dry barns? To answer these questions, we need to develop paradigms for how we should and shouldn't treat different animals. This will take time, it will take living with animals, while recognizing them as both our moral equals and as being very different from us.

A few years ago, my sister moved to Florida and rented one of the only houses on the lake without a seawall. The alligator that lived in the lake would climb into her yard and sun itself—good for the alligator, frightening for my sister. After a while, she started to get used to its presence and vowed to stop being afraid and to become friends with the alligator, which she named Gus. This presented her with a puzzle: How do we become friends with individuals we are afraid of? Ones that are dangerous for us to even get close to? How, as the writer Helen MacDonald asks, do we learn to recognize and love difference?[2] How do we at least coexist with, and even learn to appreciate, individuals who are profoundly different, without becoming intimidated by or alienated from them? Can we be friends with alligators, or do we need new models for our relationships in a world without hierarchies? Not friend, not enemy, something different without being better or worse. Eventually, Gus got too comfortable with humans, and the wildlife commission took him away, either to be killed or to be released in a remote lake. My sister was a little sad, and a lot relieved.

One might think both fates are inappropriate. Gus didn't harm anyone; hence, he didn't deserve to be killed. But he was threatening people who just wanted to be his friend. Doing that and simply being taken to another, possibly nicer home doesn't seem right either. When we threaten innocent people, we are treated very differently, and worse. We are held responsible and deemed to have done something bad, something unethical. This difference might seem unfair. Why should Gus get away with things we can't? And that sense of unfairness might seem to point to a flaw in the present analysis: How can an ostensibly ethical analysis lead to unfair outcomes?

While our superior cognitive capacities do not make us morally more important than alligators, they do endow us with the capacity to recognize and act according to the principles of morality, something Gus can't do. This difference in moral action guiding properties justifies treating humans and alligators differently, in ways that sometimes benefit us and sometimes don't. It leads to greater demands on us: we have obligations to avoid harming others and to sometimes help them, and when we fail, we face negative consequences. Our possession of superior cognitive capacities also yields significant benefits. It results, as we have seen, in our having the capacity to lead richer lives, which is good in its own right and which also makes it worse to kill us than it is to kill alligators. It also makes it more important to respect our agency.

The change that a future without moral hierarchies will bring, then, is something like the personal change I underwent several years ago while taking a walk with an elephant (see Photograph C.1: Learning to Walk with Animals). To the extent that I had thought about elephants at all, I assumed they were as lumbering cognitively as they are physically. But, even more so, I didn't really think about them at all. I assumed that human beings are significantly more important morally than elephants; hence, what they are like and how their lives should go didn't really matter, provided we don't torture them. Taking a walk with an elephant gave me a brief glimpse into what elephants are really like, and how wrong I was. I got a sense for how smart they are, how social, and how much they recognize and respond to human beings. This made me

Photograph C.1 Learning to Walk with Animals

realize that how we treat elephants can have a dramatic impact on them, and it matters morally.

For centuries, human beings have used elephants to do work for them. This sometimes requires the elephants to be chained and treated in ways that can be pretty terrible.[3] It also involves the development of relationships between elephants and their riders which become intimate and can last for decades, and to the elephants doing amazing things:

> The only work more taxing than loading logs is the clearing of logjams, which requires a level of skill and bravery possessed by few elephants indeed. The elephant must approach the logjam from downstream and search for the key log—the one whose removal will break the jam—among the hundreds that are impeding the stream. Then, with enormous strength and know-how, it must dislodge the log and get out of the way before the force of the breaching barrier harms it. It seems to me that the elephants are aware of the danger, and perhaps take pride in their work. The emotional capacities of some elephants are astonishing. Sokona was a female elephant deployed in crossing rivers. She had only recently been captured and was still only half trained when, accompanied by her calf, she was asked to cross a dangerously flooded river. Before entering the water, she grasped her calf between her tusks and the base of her trunk, her mahout and an assistant riding on her back. The violence of the torrent caused the assistant to lose his footing and plunge into the water. Sokona rescued him by grabbing him with her trunk, continuing to the far bank with both calf and human held securely in her grasp. When asked why she would rescue the man in such dangerous circumstances, her fandi (as master elephant catchers are called by the ethnic Hkamti people of northern Myanmar) said that she was "just naturally compassionate."[4]

The suggestion that elephants take pride in their work and that they are naturally compassionate struck me as fanciful, until I took a walk with one. Now, I am not surprised, and it makes me realize how rich their lives are. And it has led me to wonder whether they might have contribution interests. Was Sokona's life better for her, for her own sake, because she saved the assistant's life? I'm not sure, but I am no longer certain that it isn't.

In a way, belief in degrees of moral status had acted as a kind of blinders that kept me from seeing elephants as they really are. Instead of trying to appreciate their capacities and the nature of their subjective existences, I saw them simply as individuals who have significantly lower moral status and, hence, don't matter that much. Rejecting degrees of moral status raises the opportunity, and the challenge, of seeing them for what and who they are, and figuring out what that implies for how we ought to treat them. What types of relationships can we have with alligators and with elephants? What counts as a flourishing life for them? And how can we help them to get there?

APPENDIX

Commonly Cited Superior Cognitive Capacities

The claim that human beings have higher moral status than animals frequently is based on the fact that competent adults possess cognitive capacities which animals do not possess, at least to the same extent. These capacities are characterized in various ways, and there is great debate over which description is right. For present purposes, the details are not important, and the following descriptions are intended to be compatible with most accounts.

Consciousness and Self-Consciousness: Conscious individuals have awareness or experiences. When an autonomous vacuum runs into the sofa, it doesn't feel anything; it just moves the other way. Cats, in contrast, are conscious. When the cat runs into the sofa, it feels or experiences the contact and moves the other way. Self-conscious individuals are aware of themselves as distinct entities. When I bump into the sofa, I feel my shin throbbing, and I am aware that it is my shin that is throbbing. Some commentators argue that consciousness is necessary for having moral status and self-consciousness endows humans with greater moral status.

Narrative Identity: Individuals with a narrative identity are aware of themselves existing through time. Having a narrative identity allows individuals to plan and pursue projects and relationships. I am not aware of anyone who thinks that having a narrative identity is necessary for having moral status. But some think that having a narrative identity endows humans with greater moral status compared to animals that are sentient but do not have a narrative identity.

Agency: Trees do many things. They grow roots, and they move toward the sunlight. However, trees do not decide to do these things; they are not agents. They do not move with intentions and goals. In contrast, agents make decisions about what to do among a set of options and then try to do those things. Some people believe that being an agent is sufficient for having moral status. Others believe that it endows individuals with greater moral status.

Autonomy: Autonomous individuals (having autonomy, having the capacity to act autonomously) are self-governing agents, agents who are (or at least have the capacity to be) in charge of their lives in the sense of deciding what to do based on their views for how they want their lives to go.

Some think that being autonomous or having the capacity to be autonomous is necessary for having moral status. Others regard it as increasing individuals' level of moral status.

Moral Agency: Moral agents are agents who have the capacity to recognize moral principles and the capacity to act in light of them. The philosopher Immanuel Kant thought that being a moral agent or having the capacity for moral agency is necessary and sufficient for having moral status. Others think that moral agency increases one's moral status.

Moral Persons: Moral persons are individuals who have the capacity to recognize the principles of justice and the capacity to recognize what is good for them personally. The philosopher John Rawls believed that having these two capacities endows individuals with higher moral status.

Capacity for Mutual Engagement: Human beings have the capacity to enter into agreements with others and to live according to them. These might involve explicit agreements or tacit understanding that sharing space with others requires each to take into account the interests of others. A number of philosophers believe that this capacity is necessary and/or sufficient for having moral status.

Notes

Introduction

1. For example, a Gallup poll found that 62 percent of Americans endorsed the following claim: "Animals deserve some protection from harm and exploitation, but it is still appropriate to use them for the benefit of humans." See Rebecca Rifkin, https://news.gallup.com/poll/183275/say-animals-rights-people.aspx.
2. The example is due to Agnieszka Jaworska.
3. S. Clarke and J. Savalescu, "Rethinking Our Assumptions about Moral Status," in *Rethinking Moral Status*, ed. S. Clarke, H. Zohny, and J. Savulescu (Oxford: Oxford University Press, 2021), chap. 1, p. 12.
4. W. Gaylin, "Commentary," *Hastings Center Report*. January/February (1989): 27–28, p. 28.
5. B. Brody, "Defending Animal Research: An International Perspective," in *The Ethics of Animal Research: Exploring the Controversy*, ed. J. R. Garrett (Cambridge, MA: MIT Press, 2012), chap. 4, p. 61.
6. R. Nozick, *Anarchy, State and Utopia* (Philadelphia: Basic Books, 1974), p. 46.
7. See D. Schmidtz, "Are All Species Equal?" *Journal of Applied Philosophy* 15 (1998): 57–67.
8. See The NIH Director, "NIH Will No Longer Support Biomedical Research on Chimpanzees," November 17, 2015, https://www.nih.gov/about-nih/who-we-are/nih-director/statements/nih-will-no-longer-support-biomedical-research-chimpanzees.
9. B. M. Altevogt, D. E. Pankevich, M. E. Shelton-Davenport, and J. P. Kahn, eds. *Chimpanzees in Biomedical and Behavioral Research: Assessing the Necessity* (Washington, DC: National Academies Press, 2011).
10. Differences in degrees of moral status could also be explained by the fact that some individuals possess properties which *lower* their moral status. We will consider this possibility in section 3.2.
11. Here is a good place to start: S. Clarke, H. Zohny, and J. Savulescu, eds. *Rethinking Moral Status* (Oxford: Oxford University Press, 2021).

Chapter 1

1. H. Sidgwick, *The Methods of Ethics*, 7th ed. (Cambridge, MA: Hackett, 1981). See https://www.gutenberg.org/files/46743/46743-h/46743-h.htm.
2. T. Nagel, *The View from Nowhere* (Oxford: Oxford University Press, 1986).
3. The comparison between our views regarding the physical universe and our views regarding the moral universe is not intended to suggest that moral facts are features of the objective universe just like physical facts.
4. Here, too, there are exceptions. Followers of Jainism, an ancient religion, believe that all living things, including animals, plants, and microbes, have moral status.
5. A. Cohen, ed., *Kant's Lectures on Anthropology* (Cambridge: Cambridge University Press, 2014). Lectures on Anthropology, 7, 127. doi:10.1017/CBO9781139176170.
6. For a fascinating discussion of some of this history, and much more, see K. Thomas, *Man and the Natural World: Changing Attitudes in England 1500–1800* (Oxford: Oxford University Press, 1983).
7. For an insightful overview of this impact, see E. Mayr, *The Growth of Biological Thought: Diversity, Evolution, and Inheritance* (Cambridge, MA: Harvard University Press, 1985).
8. W. Cowper, *The Task*, https://www.gutenberg.org/files/3698/3698-h/3698-h.htm.
9. A. Trollope, *Hunting Sketches*, https://www.gutenberg.org/files/814/814-h/814-h.htm.
10. *The Guardian*, "Princely Opinions: Charles Has His Say about the Running of Britain," https://www.theguardian.com/uk/2012/oct/17/prince-charles-policy-britain-run.
11. See, for example: W. Sinnott-Armstrong and V. Conitzer, "How Much Moral Status Could AI Ever Achieve?" in *Rethinking Moral Status*, ed. S. Clarke, H. Zohny, and J. Savulescu (Oxford: Oxford University Press, 2021), chap. 16, p. 271.
12. P. Singer, *Animal Liberation*, 2nd ed. (New York: New York Review of Books, 1990), p. 8.
13. S. Kagan, *How to Count Animals, More or Less* (Oxford: Oxford University Press, 2019), p. 99.
14. S. Kagan, *How to Count Animals, More or Less* (Oxford: Oxford University Press, 2019), pp. 56–57.
15. E. Burke, *An Appeal from the New to the Old Whigs, in Consequence of Some Late Discussions in Parliament, Relative to the "Reflections on the French Revolution,"* 4th ed. (London: J. Dodsley, 1791), p. 491.

16. C. Krauthammer, "Free Willy!" *Washington Post*, May 7, 2015, http://www.washingtonpost.com/opinions/free-willy/2015/05/07/4d1a82f2-f4f2-11e4-b2f3-af5479e6bbdd_story.html.
17. The Nuremberg Code, in *Doctors of Infamy: The Story of the Nazi Medical Crimes*, ed. A. Mitscherlich and F. Mielke (New York: Schuman, 1949), pp. xxiii–xxv.
18. Such an account would need to explain why the properties that underlie our greater moral status are relevant to some of our important interests or rights, but not others.
19. T. Nagel, "What Is It Like to Be a Bat?" *The Philosophical Review* 4 (1974): 435–450.

Chapter 2

1. J. Rachels, "Drawing Lines," in *Animal Rights: Current Debates and New Directions*, ed. C. R. Sunstein and M. Nussbaum, pp. 162–174 (Oxford: Oxford University Press, 2004), p. 69.
2. Ben Sachs argues that Rachels's view suggests there is no such thing as moral status above and beyond individuals' moral action guiding properties. The present analysis largely agrees with him. But, to get there, we first need to show that there are no moral status enhancing properties. See B. Sachs, "The Status of Moral Status," *Pacific Philosophical Quarterly* 92 (2011): 87–104.
3. B. Brody, "Defending Animal Research: An International Perspective," in *The Ethics of Animal Research: Exploring the Controversy*, ed. J. R. Garrett (Cambridge, MA: MIT Press, 2012), chap. 4, p. 61.
4. A. Schweitzer, *Civilization and Ethics: The Philosophy of Civilization Part II* (London: A&C Black, 1929), p. 248.
5. See S. M. Kajiura and K. N. Holland, "Electroreception in Juvenile Scalloped Hammerhead and Sandbar Sharks," *Journal of Experimental Biology* 205 (2002): 3609–3621.
6. M. A. Warren, *Moral Status: Obligations to Persons and Other Living Things* (Oxford: Oxford University Press, 1997), chap. 6.
7. United Nations, Universal Declaration of Human Rights, 1948.

Chapter 3

1. I. Kant, *Critique of Practical Reason*, ed. L. W. Beck (New York: Liberal Arts Press, 1956).

2. See, for example: J. Waldron, *One Another's Equals: The Basis of Human Equality* (Cambridge, MA: Harvard University Press, 2017).
3. S. Kagan, *How to Count Animals, More or Less* (Oxford: Oxford University Press, 2019), p. 117.
4. N. Ferguson, *The War of the World: Twentieth-Century Conflict and the Descent of the West* (New York: Penguin Press, 2006), p. 480.
5. At least one proponent disagrees: "Some animals have less value than inanimate structures; some animals have negative value." W. Gaylin, "Commentary," *Hastings Center Report* January/February 1989: 27–28.
6. P. Fussell, *The Great War and Modern Memory* (Oxford: Oxford University Press, 1974), p. 8. For a more nuanced account of the origins of the war, see C. Clark, *The Sleepwalkers: How Europe Went to War in 1914* (New York: Harper Collins, 2013).
7. For related discussion, see S. Gosepath, "On the (Re)construction and Basic Concepts of the Morality of Equal Respect," in *Do All Persons Have Equal Moral Worth?: On 'Basic Equality' and Equal Respect and Concern*, ed. U. Steinhoff (Oxford: Oxford University Press, 2014), pp. 124–141.
8. See A. Norcross, "Animal Experimentation, Marginal Cases and the Significance of Suffering," in *The Ethics of Animal Research: Exploring the Controversy*, ed. R. Garrett (Cambridge, MA: MIT Press, 2012), chap. 5.
9. For assessment of a range of other claims that are undermined by the need for a threshold, see L. Alexander, "Scalar Properties, Binary Judgments," *Journal of Applied Philosophy* 25, no. 2 (2008): 85–104.
10. In one of the earliest discussions of this challenge, Bernard Williams argues that all human beings deserve equal treatment, given their consciousness of their place in the world and their possession of personal goals and activities. The fact that these characteristics also come in degrees raises the question of whether he solves or inherits the problem he identified. See B. Williams, "The Idea of Equality, in *Philosophy, Politics, and Society*, pp. 110–131 (London: Basil Blackwell, 1962). Also see P. Vallentyne, "Of Mice and Men: Equality and Animals," *Ethics* 9 (2005): 403–433.
11. J. Rawls, *A Theory of Justice* (Cambridge, MA: Harvard University Press, 1971), section 77.
12. J. Rawls, *A Theory of Justice* (Cambridge, MA: Harvard University Press, 1971), p. 442.
13. Christiano makes a similar point with respect to what he calls the continuity argument. T. Christiano, "Rationality, Equal Status and Egalitarianism," in *Do All Persons Have Equal Moral Worth?: On "Basic*

Equality" and Equal Respect and Concern, ed. U. Steinhoff, pp. 53–75 (Oxford: Oxford University Press, 2014), chap. 4.

14. Several proponents have attempted to identify a property which both offers a plausible basis for greater moral status and does not come in degrees. None of these attempts have been successful, however. See, for example, I. Carter, "Respect and the Basis of Equality," *Ethics* 121 (2011): 538–571; G. Sher, "Why We Are Morally Equal," in *Do All Persons Have Equal Moral Worth?*, ed. U. Steinhoff (Oxford: Oxford University Press, 2015), chap. 2. For discussion, and some of the challenges these views face, see R. Arneson, "Basic Equality: Neither Acceptable Nor Rejectable," in *Do All Persons Have Equal Moral Worth?*, ed. U. Steinhoff (Oxford: Oxford University Press, 2015), chap. 3.
15. J. Waldron, *One Another's Equals: the Basis of Human Equality* (Cambridge, MA: Harvard University Press, 2017), p. 119.
16. This would require addressing the challenge discussed earlier that there is a sense in which these capacities are both positive and negative.

Chapter 4

1. For readers interested in exploring this literature, here are three good resources: B. Bradley, F. Feldman, and J. Johansson, eds., *The Oxford Handbook of Philosophy of Death* (Oxford: Oxford University Press, 2012); S. Luper, "Death," *Stanford Encyclopedia of Philosophy*, https://plato.stanford.edu/entries/death/; and E. Gamlund and C. T. Solberg, eds., *Saving People from the Harm of Death* (Oxford: Oxford University Press, 2019).
2. See Lucretius, *On the Nature of the Universe*, trans. R. E. Latham (Hammondsworth, Middlesex: Penguin Classics, 1951).
3. Epicurus, "Letter to Menoeceus," in *Greek and Roman Philosophy after Aristotle*, ed. J. L. Saunders (New York: Free Press, 1966).
4. T. Nagel, "Death," *Nous* 4 (1970): 73–80; F. Feldman, "Some Puzzles about the Evil of Death," *The Philosophical Review* 100 (1991): 205–227.
5. Whether, in fact, killing a specific immortal jellyfish deprives it of an infinite future depends on what would have happened otherwise. If the universe will end in five billion years, as one theory predicts so, too, will all immortal jellyfish.
6. If immortal jellyfish also have some basic psychological connection to their future pleasures, and they are potentially infinite in number, their deaths will be worse than ours on a time relative interests account (TRIA) as well.

7. J. Glover, *Choosing Children: The Ethical Dilemmas of Genetic Intervention* (Oxford: Clarendon Press, 2006), chap. 3. For an account of what makes achievements valuable and an argument that doing something well can be valuable, even when the activity itself has no intrinsic value, see T. Hurka and J. Tasioulas, "Games and the Good," *Proceedings of the Aristotelian Society* 80 (2006): 217–264.
8. Determining whether contributing to a specific valuable project promotes individuals' interests requires answering a number of questions: Do the individuals have to know they made the contribution? Do they have to make the contribution intentionally? Do the individuals have to regard the project as valuable? Do they have to contribute to the project because they regard it as valuable? For discussion of these questions, see D. Wendler, *The Ethics of Pediatric Research* (Oxford: Oxford University Press, 2010), chaps. 6–8.
9. Compare: "To kill a person is therefore, normally, to violate not just one, but a wide range of the most central and significant preferences a being can have. Very often, it will make nonsense of everything that the victim has been trying to do in the past days, months, or even years." P. Singer, *Practical Ethics* (Cambridge: Cambridge University Press, 1981), p. 95.
10. This claim has been endorsed by many philosophers, leading to a voluminous literature and numerous variations. See, for example, D. DeGrazia, *Taking Animals Seriously: Mental Life and Moral Status* (Cambridge: Cambridge University Press, 1996), chap. 8.
11. This is similar to, but narrower than Martha Nussbaum's view regarding the badness of death: "When a life contains a temporal unfolding of which the individual subject is aware and which the subject values, death can harm it." As Nussbaum notes, reading a long novel is an activity which we pursue over time. On her view, the fact that one's death cuts off that activity makes death bad for the individual. On my view, that depends on the extent to which reading and finishing the novel qualifies as a valuable project. See: M.C. Nussbaum, Justice for Animals: Our Collective Responsibility (New York: Simon and Schuster, 2022), pp. 160–161.
12. For similar thoughts on what distinguishes the deaths of human beings and animals, see J. McMahan, *The Ethics of Killing: Problems at the Margins of Life* (Oxford: Oxford University Press, 2002), especially pp. 197–199. For some doubts, and an alternative account, see J. Shepherd, *Consciousness and Moral Status* (London: Routledge, 2018).
13. For one attempt to characterize this value, see R. Dworkin, *Life's Dominion* (New York: Alfred Knopf, 1993), pp. 70–85.

14. J. McMahan, "Challenges to Human Equality," *Ethics* 12 (2008): 81–104, p. 82. McMahan is discussing persons, not human beings. There is extensive literature evaluating this view. For example: K. Lippert-Rasmussen, "Why Killing Some People Is More Seriously Wrong Than Killing Others," *Ethics* 117, no. 4 (2007): 716–738; C. Soto, "Killing, Wrongness, and Equality," *Philosophical Studies* 164, no. 2 (2013): 543–559; M. Hanser, "The Wrongness of Killing and the Badness of Death," in *The Oxford Handbook of Philosophy of Death*, ed. B. Bradley, F. Feldman, and J. Johansson (Oxford: Oxford University Press, 2012), chap. 17.

Chapter 5

1. These differences, and their implications, are explored in B. Rollins, *The Unheeded Cry: Animal Consciousness, Animal Pain, and Science*, 2nd ed. (Ames: Iowa State University Press, 1998).
2. This should be distinguished from the possibility that the less developed cognitive capacities of nonhuman animals might result in their having lower levels of consciousness. See D. Griffin, *Animal Minds: Beyond Cognition to Consciousness* (Chicago: University of Chicago Press, 2001), especially p. 18.
3. For more on the possibility of an anti-animal bias, see D. DeGrazia, *Taking Animals Seriously: Mental Life and Moral Status* (Cambridge: Cambridge University Press, 1996), chap. 3; and E. B. Pluhar, *Beyond Prejudice: The Moral Significance of Human and Nonhuman Animals* (Durham, NC: Duke University Press, 1995).
4. See, for example: N. Haslam and P. Bain, "Humanizing the Self: Moderators of the Attribution of Lesser Humanness to Others," *Personality and Social Psychology Bulletin* 33 (2007): 57–68, doi:10.1177/0146167206293191; and N. Haslam, P. Bain, L. Douge, M. Lee, and B. Bastian, "More Human Than You: Attributing Humanness to Self and Others," *Journal of Personality and Social Psychology* 89 (2005): 937–950, doi:10.1037/00223514.89.6.937.
5. B. Bastian, S. M. Laham, S. Wilson, N. Haslam, and P. Koval, "Blaming, Praising, and Protecting Our Humanity: The Implications of Everyday Dehumanization for Judgments of Moral Status," *British Journal of Social Psychology* 50 (2011): 469–483.
6. B. Bastian, S. M. Laham, S. Wilson, N. Haslam, and P. Koval, "Blaming, Praising, and Protecting Our Humanity: The Implications of Everyday Dehumanization for Judgments of Moral Status," *British Journal of Social Psychology* 50 (2011): 469–483.

7. J. Hecht and A. Horowitz, "Seeing Dogs: Human Preferences for Dog Physical Attributes," *Anthrozoos: A Multidisciplinary Journal of the Interactions of People & Animals* 28, no. 1 (2015): 153–163. doi:10.2752/089279315X14129350722217.

Chapter 6

1. *Doe ex. rel. Tarlow v. District of Columbia*, 489 F.3d 376 (D.C. Cir. 2007).
2. Specifically, it would save 3,750,000 years of life, every year. For explanation of the numbers and elaboration of the argument, see D. Wendler, "A Test of Utilitarianism for Animals, Kantianism for People," *Journal of Moral Philosophy* 18 (2021): 473–499.
3. Christine Korsgaard defines agency as "representation-governed locomotion." C. M. Korsgaard, *Fellow Creatures: Our Obligations to the Other Animals* (Oxford: Oxford University Press, 2018), p. 23. The present understanding is even broader. An individual that randomly searches its environment for food, without any representation of it, would qualify as an agent on the present account, but not on her account.
4. World Medical Association, "Declaration of Helsinki: Ethical Principles for Medical Research Involving Human Subjects," *JAMA* 310, no. 20 (2013): 2191–2194, https://doi.org/10.1001/jama.2013.281053.
5. For the full argument, see D. Howard and D. Wendler, "Beyond Instrumental Value: Respecting the Will of Others and Deciding on Their Behalf," in *The Oxford Handbook of Philosophy and Disability*, ed. A. Cureton and D. Wasserman (Oxford: Oxford University Press, 2020), chap. 30.
6. S. Shiffrin, "Paternalism, Unconscionability Doctrine, and Accommodation," *Philosophy & Public Affairs* 29 (2000): 205–250, pp. 218–219.
7. For an argument that, in a fundamental sense, there are no subjective experiences or consciousness, see D. C. Dennett, *From Bacteria to Bach and Back: The Evolution of Minds* (New York: WW Norton, 2017). For a review, and healthy dose of skepticism, see T. Nagel, "Is Consciousness an Illusion?" *New York Review of Books*, March 9, 2017.
8. See, for example: C. Safina, *Beyond Words: What Animals Think and Feel* (New York: Henry Holt and Co, 2015); and F. B. M. de Waal and P. L. Tyack, eds., *Animal Social Complexity: Intelligence, Culture, and Individualized Societies* (Cambridge, MA: Harvard University Press, 2003).
9. C. Sueur and O. Petit, "Shared or Unshared Consensus Decision in Macaques?" *Behavioural Processes* 78 (2008): 84–92.

10. C.A. H. Bousquet, D. J. T. Sumpter, and M. B. Manser, "Moving Calls: A Vocal Mechanism Underlying Quorum Decisions in Cohesive Groups," *Proceedings of the Royal Society B* 248 (2011): 1482–1488.
11. R. H. Walker, A. J. King, A. W. McNutt, and N. R. Jordan, "Sneeze to Leave: African Wild Dogs (*Lycaon pictus*) Use Variable Quorum Thresholds Facilitated by Sneezes in Collective Decisions," *Proceedings of the Royal Society B* 284, no. 1862 (2017): 20170347. doi:10.1098/rspb.2017.0347. PMID: 28878054; PMCID: PMC5597819.
12. M. A. van Noordwijk, "From Maternal Investment to Lifetime Maternal Care," *The Evolution of Primate Societies*, ed. J. C. Mitani, J. Call, P. M. Kappeler, R. A. Polombit, and J. B. Silk (Chicago: University of Chicago Press, 2012), pp. 321–342.
13. See, for example: J. Sebo, "Agency and Moral Status," *Journal of Moral Philosophy* 14, no. 1 (2017): 1–22.

Chapter 7

1. T. M. Scanlon, *What We Owe to Each Other* (Cambridge, MA: Harvard University Press, 1998), p. 185.
2. A different view emphasizes the full and equal moral status of all humans based, not on our species, but on our human form. See R. Kipke, "Being Human: Why and in What Sense It Is Morally Relevant," *Bioethics* 34, no. 2 (2020): 148–158.
3. For a related argument, see C. Diamond, "Eating Meat and Eating People," *Philosophy* 53, no. 206 (1978): 465–479.
4. B. Brody, "Defending Animal Research: An International Perspective," in *The Ethics of Animal Research: Exploring the Controversy*, ed. J. R. Garrett (Cambridge, MA: MIT Press, 2012), chap. 4, p. 62.
5. See, for example, S. Keller, *Partiality* (Princeton, NJ: Princeton University Press, 2013); N. W. Olson, "Review: Simon Keller Partiality," *Ethics* 124, no. 3 (2014): 622–626; J. Raz, "Liberating Duties," *Law and Philosophy* 8, no. 1 (1989): 3–21; S. Scheffler, *Boundaries and Allegiances: Problems of Justice and Responsibility in Liberal Thought* (Oxford: Oxford University Press, 2003).
6. H. Brighouse and A. Swift, "Legitimate Parental Partiality," *Philosophy & Public Affairs* 37, no. 1 (2009): 43–80; H. Brighouse and A. Swift, "Parents' Rights and the Value of the Family," *Ethics* 117 (2006): 80–108.
7. C. Palmer, *Animal Ethics in Context* (New York: Columbia University Press, 2010), p. 77.

8. L. P. Francis and R. Norman, "Some Animals Are More Equal Than Others," *Philosophy* 53 (1978): 507–527.
9. C. Palmer, *Animal Ethics in Context* (New York: Columbia University Press, 2010), p. 123.
10. B. Williams, "The Human Prejudice," in *Philosophy as a Humanistic Discipline*, pp. 135–152 (Princeton, NJ: Princeton University Press, 2006).
11. For extended discussion of this and related issues, see A. Buchanan, *Our Moral Fate: Evolution and the Escape from Tribalism* (Cambridge, MA: MIT Press, 2020).

Chapter 8

1. For discussion and possible answers, see K. Mack, *The End of Everything (Astrophysically Speaking)* (New York: Scribner, 2020).
2. National Commission for the Protection of Human Subjects of Biomedical and Behavioral Research, "The Belmont Report: Ethical Principles and Guidelines for the Protection of Human Subjects of Research, 1978, section 3," https://www.hhs.gov/ohrp/regulations-and-policy/belmont-report/read-the-belmont-report/index.html#xselect.
3. S. Kagan, *How to Count Animals, More or Less* (Oxford: Oxford University Press, 2019), p. 48.
4. World Health Organization, "Global Spending on Health: Weathering the Storm," https://www.who.int/publications/i/item/9789240017788.
5. Vallentyne endorses this response. See P. Vallentyne, "Of Mice and Men: Equality and Animals," *Ethics* 9 (2005): 403–433.
6. Most prominently, see S. Kagan, *How to Count Animals, More or Less* (Oxford: Oxford University Press, 2019), chap. 3.
7. For a moving account of these cases, and their impact, see A. Gregory, "The Sorrow and the Shame of the Accidental Killer," *The New Yorker*, September 11, 2017.
8. See, for example, T. Regan, *The Case for Animal Rights* (Berkeley: University of California Press, 2004).
9. S. Donaldson and W. Kymlicka, *Zoopolis: A Political Theory of Animal Rights* (Oxford: Oxford University Press, 2011).
10. E. Pluhar, "Experimentation on Humans and Nonhumans," *Theoretical Medicine and Bioethics* 27 (2006): 333–355, p. 335.
11. This view is similar to what James Rachels calls moral individualism. J. Rachels, *Created from Animals: The Moral Implications of Darwinism* (Oxford: Oxford University Press, 1990), chap. 5. Rachels seems to think

that individual properties are all that matter to determining how we ought to treat others. I think, in contrast, that the type to which an individual belongs is critical for determining what constitutes a flourishing life for them and, hence, how they ought to be treated.

12. T. Regan, *The Case for Animal Rights*, 2nd ed. (Berkeley: University of California Press, 2004), pp. 237–243. This approach is endorsed by a range of authors. For example, Donaldson and Kymlicka argue that our inviolable rights are grounded in our selfhood, which we share with animals, suggesting that they have the same rights. They argue that other attempts to ground our rights will inevitably leave out children and humans with significant cognitive disabilities. See S. Donaldson and W. Kymlicka, *Zoopolis: A Political Theory of Animal Rights* (Oxford: Oxford University Press, 2011), pp. 40–45.
13. M. L. Wilson, Michael L. Wilson, Christophe Boesch, Barbara Fruth, Takeshi Furuichi, Ian C. Gilby, Chie Hashimoto, Catherine L. Hobaiter, Gottfried Hohmann, Noriko Itoh, Kathelijne Koops, Julia N. Lloyd, Tetsuro Matsuzawa, John C. Mitani, Deus C. Mjungu, David Morgan, Martin N. Muller, Roger Mundry, Michio Nakamura, Jill Pruetz, Anne E. Pusey, Julia Riedel, Crickette Sanz, Anne M. Schel, Nicole Simmons, Michel Waller, David P. Watts, Frances White, Roman M. Wittig, Klaus Zuberbühler, and Richard W. Wrangham, "Lethal Aggression in Pan Is Better Explained by Adaptive Strategies Than Human Impacts," *Nature* 513 (2014): 414–417, doi:10.1038/nature13727.
14. M. C. Nussbaum, *Justice for Animals: Our Collective Responsibility* (New York: Simon and Schuster, 2022), pp. 160–161. Also see M. C. Nussbaum, "What We Owe Our Fellow Animals," *New York Review of Books* LXIX, no. 4 (2022): 34–36.
15. Maryland Department of Natural Resources, "Smallwood State Park," https://dnr.maryland.gov/publiclands/pages/southern/smallwood.aspx.
16. See "The Moral Problem of Predation," in *Philosophy Comes to Dinner Arguments About the Ethics of Eating*, ed. A. Chignell, T. Cuneo, and M. C. Halteman (New York: Routledge, 2015), pp. 268–294.
17. C. M. Korsgaard, *Fellow Creatures: Our Obligations to the Other Animals* (Oxford: Oxford University Press, 2018), section 4.3.8.
18. E. B. Pluhar, *Beyond Prejudice: The Moral Significance of Human and Nonhuman Animals* (Durham, NC: Duke University Press, 1995), p. 292.
19. See, for example, S. Adamo, "Is It Pain If It Does Not Hurt? On the Unlikelihood of Insect Pain," *The Canadian Entomologist* 151, no. 6 (2019): 685–695, doi:10.4039/tce.2019.49.

20. H. Ritchie and M. Roser, "Meat and Dairy Production," in *Our World in Data* (Oxford: Oxford University Press, 2019), https://ourworldindata.org/meat-production.
21. See J. S. Foer, "The End of Meat Is Here," https://www.nytimes.com/2020/05/21/opinion/coronavirus-meat-vegetarianism.html?smid=nytcore-ios-share.
22. For example, Tom Regan argues that humane raising and slaughtering of animals harms them because "All that animal's future prospects for satisfaction are denied." T. Regan, *The Case for Animal Rights*, 2nd ed. (Berkeley: University of California Press, 2004), p. 336.
23. Even if one insists that individuals' experiential interests are not contingent in the way I have argued, humane animal farming may still turn out to be permissible. If we stop animal farming, the animals that would have been raised are deprived of the pleasures they would have enjoyed. See, for example, C. Solis, "How Much Does Slaughter Harm Humanely Raised Animals?," *Journal of Applied Philosophy* 38, no. 2 (2021): 258–272.
24. W. M. S. Russell and R. L. Burch, *The Principles of Humane Experimental Technique* (Wheathampstead, UK: Universities Federation for Animal Welfare, 1959; reprinted in 1992).
25. C. Natanson, R. L. Danner, M. P. Fink, T. J. MacVittie, R. I. Walker, J. J. Conklin, and J. E. Parrillo, "Cardiovascular Performance with E. coli Challenges in a Canine Model of Human Sepsis," *American Journal of Physiology: Heart and Circulatory Physiology* 254 (1988): H558–H569.
26. A number of recent works have argued for changes to animal research regulations. T. L. Beauchamp and D. DeGrazia, *Principles of Animal Research Ethics* (Oxford: Oxford University Press, 2020); H. Ferdowsian, L. S. M. Johnson, J. Johnson, A. Fenton, A. Shriver, and J. Gluck, "A Belmont Report for Animals?," *Cambridge Quarterly Healthcare Ethics* 29, no. 1 (2020): 19–37, doi:10.1017/S0963180119000732. For a brief overview, see G. Arnason, "The Emergence and Development of Animal Research Ethics: A Review with a Focus on Nonhuman Primates," *Science and Engineering Ethics* 26, no. 4 (2020): 2277–2293, doi:10.1007/s11948-020-00219-z.
27. US Code of Federal Regulations, 45CFR46.404/50.51.
28. US Code of Federal Regulations, 45CFR46.407/50.54.
29. Institute of Medicine, *Science, Medicine, and Animals* (Washington, DC: The National Academies Press, 1991), p. 4. https://nap.nationalacademies.org/catalog/10089/science-medicine-and-animals.

30. Researchers are learning to determine whether and to what extent individuals are in pain by scanning their brains. See, for example, T. D. Wager, L. Y. Atlas, M. A. Lindquist, M. Roy, C. W. Woo, and E. Kross, "An fMRI-Based Neurologic Signature of Physical Pain," *New England Journal of Medicine* 368, no. 15 (2013): 1388–1397. This approach works in humans because the investigators can determine what a certain pattern of brain activity feels like, how much it hurts, by asking individuals. For an overview, see I. Tracey, "Finding the Hurt in Pain," *Cerebrum* Nov–Dec (2016): 15–16.
31. D. Wendler, E. Abdoler, L. Wiener, and C. Grady, "Views of Adolescents and Parents on Pediatric Research Without the Potential for Clinical Benefit," *Pediatrics* 130, no. 4 (2012): 692–699, doi:10.1542/peds.2012-0068.
32. For further analysis and discussion, see D. Wendler, "Suffering in Animal Research: The Need for Limits and the Possibility of Compensation," *Kennedy Institute of Ethics Journal* 32, no. 3 (2022): 297–311. doi:10.1353/ken.2022.0019..
33. For some discussion of these studies, and the associated ethical issues, see H. Kantin and D. Wendler, "Is There a Role for Assent or Dissent in Animal Research?," *Cambridge Quarterly of Healthcare Ethics* 24 (2015): 459–472.
34. For a compelling account of the potential impact, see J. P. Gluck, *Voracious Science and Vulnerable Animals: A Primate Scientist's Ethical Journey* (Chicago: University of Chicago Press, 2016).
35. Kieran Setiya tries to rescue humanism from specieism by arguing that being human (as distinct from the biological definitions of *Homo sapiens* in terms of geneology, genes, or interbreeding) gives individuals special moral status on the grounds that conforming to or recognizing the standards of ethics is part of the proper functioning and, therefore, flourishing life, of human beings. See K. Setiya, "Humanism," *Journal of the American Philosophical Association* (2018): 452–470.
36. Most prominently, see E. Harman, "The Ever Conscious View and the Contingency of Moral Status," in *Rethinking Moral Status*, ed. S. Clarke, H. Zohny, and J. Savulescu (Oxford: Oxford University Press, 2021), chap. 6.
37. E. Kittay, "At the Margins of Moral Personhood," *Ethics* 116, no. 1 (2005): 100–131.
38. C. Cohen, "The Case for the Use of Animals in Biomedical Research," in *Contemporary Moral Issues: Diversity and Consensus*, 4th ed., ed. L. M. Hinman (London: Routledge, 2013), chap. 10.
39. J. Waldron, *One Another's Equals: The Basis of Human Equality* (Cambridge, MA: Harvard University Press, 2017).

40. Shelly Kagan tries to address this challenge by defending modal personism. See S. Kagan, "What's Wrong with Speciesism?," *Journal of Applied Philosophy* 33, no. 1 (2016): 1–21.
41. For brief discussion of the concerns this raises, see A. Regalado, "Engineering the Perfect Baby: Scientists Are Developing Ways to Edit the DNA of Tomorrow's Children. Should They Stop Before It's Too Late?," *MIT Technology Review*, https://www.technologyreview.com/s/535661/engineering-the-perfect-baby/.
42. D. DeGrazia, "Genetic Enhancement, Post-Persons and Moral Status: A Reply to Buchanan," *Journal of Medical Ethics* 38, no. 3 (2012): 135–139, doi:10.1136/medethics-2011-100126.
43. See, for example, F. J. Zachariah, L. A. Rossi, L. M. Roberts, and L. D. Bosserman, "Prospective Comparison of Medical Oncologists and a Machine Learning Model to Predict 3-Month Mortality in Patients with Metastatic Solid Tumors," *JAMA Network Open* 5, no. 5 (2022): e2214514, doi:10.1001/jamanetworkopen.2022.14514.
44. See S. Dehaene, H. Lau, and S. Kouider, "What Is Consciousness, and Could Machines Have It?," *Science* 358, no. 6362 (2017): 486–492, doi:10.1126/science.aan8871. For discussion of the fact that we might be getting close to sentient artificial intelligence, or might already be there, see N. Tiku, "The Google Engineer Who Thinks the Company's AI Has Come to Life," *Washington Post*, June 11, 2022, https://www.washingtonpost.com/technology/2022/06/11/google-ai-lamda-blake-lemoine/.
45. For discussion of the potential and the pitfalls, see N. Bostrom, *Superintelligence: Paths, Dangers, Strategies* (Oxford: Oxford University Press, 2014).
46. R. J. Arneson, "Equality," in *The Blackwell Guide to Social and Political Philosophy*, ed. R. L. Simon (Oxford: Blackwell, 2002), chap. 4.
47. See, for example, A. Buchanan, "Moral Status and Human Enhancement," *Philosophy & Public Affairs* 37, no. 4 (2009): 346–381, especially pp. 364–367.
48. For brief discussion and relevant references, see R. R. Faden, T. L. Beauchamp, D. J. H. Mathews, and A. Regenberg, "Toward a Theory of Moral Status Inclusive of Nonhuman Animals," in *Rethinking Moral Status*, ed. S. Clarke, H. Zohny, and J. Savulescu (Oxford: Oxford University Press, 2021), chap. 10.

Conclusion

1. There has been a lot of recent work on the psychological and evolutionary bases for our beliefs regarding moral status. One question concerns whether our tendency to favor humans over animals has an evolutionary basis and, if so, whether we might be able to overcome it. See, for example, T. J. Kasperbauer, *Subhuman: The Moral Psychology of Human Attitudes to Animals* (Oxford: Oxford University Press, 2018).
2. H. Macdonald, *Vesper Flights* (New York: Grove Atlantic, 2020).
3. For discussion of how humans interact with elephants, and what is involved, for both good and ill, see J. Shell, *Giants of the Monsoon Forest: Living and Working with Elephants* (New York: WW Norton, 2019).
4. T. Flannery, "Man's Biggest Friend," *New York Review of Books*, November 21, 2019.

Conclusion

Index

For the benefit of digital users, indexed terms that span two pages (e.g., 52–53) may, on occasion, appear on only one of those pages.
Figures and boxes are indicated by *f* and *b* following the page number